*Pondering Our Existence
for Practical Living*

Pondering Our Existence for Practical Living

Evidence Analyzed by Science and Logic

Stephen A. Gentile

Solofrani Press

Pondering Our Existence for Practical Living
Evidence Analyzed by Science and Logic

Author: Stephen Gentile

Front Cover by Elizabeth Gentile

Copyright © 2025 by Stephen Gentile
All Rights reserved

Printed in the United States of America
Thank you for complying with copyright rules, no part of this publication may be reproduced, scanned, or distributed, or transmitted in any form by means electronic, mechanical in any form without permission.

ISBN: 979-8-218-67665-0
Solofrani Press

Acknowledgement

To analyze and answer the questions, "Why are we here?" "Why do I exist?" has massive impact on our lives. In undertaking this endeavor, it is evident that there are obvious paradoxes. I have strong peer-reviewed research (nearly 200 referenced, peer-reviewed, citations) for all points made in this book, even points that seem to be in conflict. There are sections in this book that come from discussions with my son, Daniel Gentile. Daniel is analytical, he's an award winning, innovative, computer engineer. He contributed to the ideas in this book on design emerging from simple rules. His analysis of π was another vital contribution. There are many possibilities put forth in this work. I hope to improve lives with these ideas, which are based in what we see naturally, scientifically, and logically.

We are all going to die one day, so let's live our lives the best way we can find, the way we want to.

Table of Contents

CHAPTER 1 1

    PURPOSE OF THIS BOOK,
    WEIGHING OPTIONS

CHAPTER 2 6

    BE OPEN TO NEW IDEAS
    ABOUT REALITY AND TRUTH

CHAPTER 3 8

    CURRENT SCIENTIFIC KNOWLEDGE
    FREE WILL/INDETERMINATION IN QUANTUM PHYSICS.

CHAPTER 4 13

    MIND CLOUD, UNIVERSAL CONSCIOUSNESS, AND EXTERNAL BRAIN WAVES

CHAPTER 5 22

    TIME & SPACE

CHAPTER 6 27

    DESIGN, CHAOS AND INFINITE POSSIBILITIES

CHAPTER 7 33

    SCIENCE & LOGIC GUIDE OUR UNDERSTANDING, EMOTION IS POWERFUL
    A WORD ABOUT MUSIC

CHAPTER 8 36

    DESIGN, EQUILIBRIUM,
    CHAOS AND ORDER

CHAPTER 9 41

    NIHILISM V PURPOSE DRIVEN, DESIGN FAVORS PURPOSE

CHAPTER 10 .... **48**

REASONS FOR PURPOSE OVER NIHILISM

CHAPTER 11 .... **52**

CREATORS, OTHER POSSIBILITIES & FUNCTIONALITY

CHAPTER 12 .... **56**

DO GOD(S) INTERVENE?

CHAPTER 13 .... **62**

WHAT CONSTITUTES A WELL-LIVED LIFE? IF THERE IS A PURPOSE, WHAT IS IT?

CHAPTER 14 .... **68**

PICKING A SECONDARY PURPOSE, BEYOND AUTOMATIC/NATURAL PURPOSE

CHAPTER 15 .... **75**

HOW WE DECIDE WHAT IS TRUTH

CHAPTER 16 .... **87**

UBIQUITOUS CONTINUUM #1:
GOOD/BAD
UNIVERSAL PULL OF RIGHT/WRONG
WITH AFTERLIFE CONSIDERATIONS

CHAPTER 17 .... **100**

THE VIOLENCE COMPASSION CONTINUUM

CHAPTER 18 .... **103**

AFTERLIFE IDEAS, FAITH SHIFTS BLAME

CHAPTER 19 .... **107**

AFTERLIFE IDEAS: REINCARNATION

CHAPTER 20 111

    UBIQUITOUS CONTINUUM #2:
    CONSTRUCTIVE/DESTRUCTIVE
    MURDEROUS AGGRESSION &
    CARING COMPASSION

CHAPTER 21 126

    UBIQUITOUS CONTINUUM #3:
    FREEDOM/RESTRICTION
    BASIC RIGHTS TO YOUR OWN BODY

CHAPTER 22 131

    HIERARCHY OF VIRTUES
    INTELLECT, CONFIDENCE, SELF-CONTROL
    AND PHYSICAL PROWESS

CHAPTER 23 145

    INTERPERSONAL RELATIONS,
    COMMUNICATION & FRIENDSHIP

CHAPTER 24 149

    UBIQUITOUS CONTINUUM #4:
    MAGNANIMITY/SELFISHNESS

CHAPTER 25 153

    RELIGIONS AND GOVERNMENTS
    DETER FREEDOM

BIBLIOGRAPHY 172

AUTHOR BIOGRAPHY 200

Chapter 1

Purpose of this Book, Weighing Options

Why are we here? Why are we alive? How do we even know "here" truly exists? Is this life just a big joke? Is there a best way for people to have the most fulfilling lives? Questioning, searching for answers, this is a quest to put forth observations, clarify the different possibilities, and help pick a path by which to live. Where do our morals, and our actions (steered by our morals) come from? Is there an evolutionary benefit? Is there a truth to be understood? What we call "truth" is certainly impacted by our own viewpoint; truth isn't the truth to the unlearned. What we call truth often looks false to those who cannot fathom change, especially when they are confronted with new knowledge. There are many people and institutions who will gladly attempt to brainwash/indoctrinate others as to the reason for our existence. There are historical records of people who were supposedly close to God, or creators, that are taught to us. This indoctrination starts at birth. Let us step back, and truly weigh all the possibilities. We are all going to die one day, this book takes on the 'how' and 'why' of life, which impacts the way we live our lives. This, in turn, has ramifications on our society, our culture, and our political setup.

One must exist in relative safety, with basic needs of food and shelter met, to be able to dwell on philosophical thought. It is only possible to ponder existence when we have time and safety. It is most effective to use our academic knowledge and our rational, ability to reason, when we decide on any belief, or course of action. This is what constitutes what is considered most likely true throughout this book: science, and reason. Interestingly, our emotional self holds the most power when we decide on any belief. We could be influenced by any mix of stories, cultural pressure, or research studies. In the end, if we're honest with ourselves, we seem to have decision making ability which is guided by our internal, emotional pull. Our internal feelings about what we know, this is what guides our life choices. The meaning and possible ramifications of what people label "truth" is a difficult mountain to climb. Commonly (and sadly) often times "truth" just means... "I agree with what is being presented, because it coincides with my currently held beliefs." So, if you read this book with the idea of bolstering your already held views of this existence, I imagine this information may, at times, (hopefully!) challenge and provoke new ideas.

This writing is an alternative approach to the prevalent tactic of swallowing ideas and information trumpeted by governments, prophets, parents, gurus, teachers and advertisements. Let us openly look at options and possibilities. Let's use 1. Reason/rationality, 2. scientific evidence, and 3. adherence paid to gut, emotional reaction. These three main areas are chosen over the other ways to determine why we might exist, because they're based in reality, experience, with science evidence backing the ideas. As we live our lives, especially as we age, we see people who are close to us dying. My mother and father were amazing parents. When they died, I had thoughts of, "Is this life just a cruel, random joke?" Yet, equally true, we feel injustice, we long for reason...

why? This book focuses on figuring out the possibilities of what is most likely true, in the end, we all choose for ourselves. How we answer these deep questions about existence determines how we live our lives, and guides the decisions we make when speaking and choosing our actions moment by moment. How should we live as individuals? How do we exist as a part of society, and our environment on the whole?

The ideas in this book are merged together from many sources: philosophers, scientists, logical reasoning, and experiences in counseling and dealing with thousands of people throughout my working life. For the past twenty-eight years, my job has allowed me to teach and counsel through an inordinate number of interpersonal situations. People who were enduring some of the most grueling situations you can imagine. This has forced me to question everything, while examining human situations and viewpoints. Through these thousands of discussions, my hope is to pass on ideas as to why we are here, and what is meant by living effectively. Some of these ideas, anyone with philosophical and/or scientific backgrounds will say, "That idea came from (Fill in the blank)!" Yes, definitely, but the value of the following is that I synthesize empirical evidence, physics, and reason, to come to a better picture of reality. The goal is to link, "Why are we here," to what we observe naturally, and scientifically. I hope to change our world for the better, or at least try to. Mother Teresa was mocked for only helping the people around her, not impacting the world in a "big way." She responded, "Never worry about numbers. Help one person at a time and always start with the person nearest to you" (Surles, 2016).

One point is certain; our existence does not have to be limited to the direction of a few leaders in religious or governmental roles. The people "at the top," do not have esoteric knowledge or special power, yet they do influence, indoctrinate, and manipulate the masses. Let us focus on agreed upon, observable evidence, and this is what should be considered truth. A belief in some fact occurring in our perceived existence, real to us, our real-world experiences. We should base our lives on this as being the closest we can get to truth. Especially when compared to accepting others' stories. Surely, many people base their entire lives on false premises. Many would say truth is in observable, shared, agreed upon "facts" in objective reality… science. Many believe science is absolute truth, but "science" has been untruthful many times throughout history. In fact, Kurt Gödel (20th century mathematician) proved that within a mathematical system, there are true statements which cannot be proven (Godel, 1999; Schmidhuber, 2009). Every mathematical and scientific proof has assumptions and limitations. For this book, truth is best defined as being observable, empirical evidence, with use of reason and logic, and even involves listening to our emotional reaction to the important ideas about existence. No beliefs, no ideas, should ever be "off the table." Your conclusions are as valid as mine, or anybody else's. Creation of your own reality, ideas about the afterlife, what constitutes "beneficial," or "harmful," these are just a few of the variable parts of our lives. Let's consider all options, and then individually decide what seem to be the most likely, valid answers.

The effect of our physical situation, the earth's position in the universe, its rotation and revolution, this makes us age. There is a physical beginning, we are existing in this perceived space-time realm where we live, and then, we physically die. We complete actions in order to survive, we complete actions to please our senses, we complete actions to hurt and/or help others. We pick actions to undertake that we think are important. Hopefully, we have the opportunity to do the things we think are important, or bring pleasure to our five senses. Many people have existed under times of heavy government, or religious control, and may have had little to no opportunity to complete actions that would bring tranquility, or deep meaning to their existence. What is surprising is how many people are willing give their thoughts, actions and time… their very lives away! First, comes the acceptance of others' beliefs, others' thoughts, and others' opinions. Then, these beliefs, thoughts and opinions lead to actions, actions that take up vast quantities of time in our individual lives. Be willing to question, to make progress. Possibly, the ideas presented will have a grand impact, an immense amount of people that would throw off the shackles of superstition, obvious error, and traditional/dominant thought, and to actually take a step closer to reality. Individual self-inspection will push society closer to an improved existence. The evidence and ideas in this book will hopefully provoke you to ponder your own, chosen, best options.

Chapter 2

Be Open to New Ideas About Reality and Truth

For twenty years I was a surveyor for the United States Forest Service. Where the boundary line went, was where our survey crew had to go. Sometimes the boundary line went in very remote places, even steep cliffs! Once, I was on top of a remote mountain, very high elevation, in the Targhee National Forest. There was nothing but pure nature around for many miles, no lights, no road, only the occasional animal, maybe, would cross this remote mountaintop. I flipped over a rock on top of this mountain, and sure enough, there was an ant colony living their lives under this rock. Just think about the entire universe surrounding this little ant colony - all the space, the air, trees, planets, all objects, animals, the unseen subatomic world… the infinite universe. Let's say these ants have knowledge, feelings, and a brain. The surrounding environment of which these ants have interactions is comparatively, severely limited. Everything in the universe surrounding this remote ant colony represents knowledge that could be studied by these isolated ants. The ants, trapped at the rocky mountaintop have such a limited chance of understanding anything of this vast universe.

Pondering Our Existence for Practical Living

The closed rock home limits any possibility of the ants understanding reality. Indeed, even seeing the truth of his/her own ant-existence. The ants may have thorough knowledge of their own isolated environment, but do the ants know Paris exists? Do the ants know they are made of atoms? Do they know what a human is? The weird creature that just flipped over their rock-covered home? These ants are a comparison to us humans. This is how most humans live our entire lives – blindly and robotically. As babies are often immediately sworn into the "mountaintop rock" of a limited, illusory meaning for existence. This is what people do when they swallow others' ideas without any critical thought or analysis. The ants on the remote mountain top are a metaphor for the people who choose the path of auto-pilot/accepting, believing, and devoting their lives to the rock walls that block their intellectual view.

Chapter 3

Current Scientific Knowledge Free Will/Indetermination in Quantum Physics.

Our human emotions, personality, and decisions, some would say, could all resolve down to chemistry. We are made of physical particles which follow predefined laws that are out of our control. However, new discoveries of quantum probabilities and quantum randomness allows different variables to get plugged into fixed rules and generates... sometimes... unexpected behavior. Quantum mechanics has shown that every particle in the universe has a wave function, particles in that wave are said to be in "superposition." Every subatomic particle is in separate states, at the same moment. The best humans can do is predict the probability of seeing a particle (i.e. an electron, or photon, etc.) in any particular location, because these particles are always in waves. A radioactive atom is decaying, or, it is not decayed. A person did drink a glass of water, or, did not drink a glass of water. Erwin Schrödinger's famous cat-in-a-box thought experiment... the unseen cat in a box is dead, or, the unseen cat is alive. Molecules, photons, even humans, and entire systems, have possibilities to be in different states.

Pondering Our Existence for Practical Living

This is simultaneously weird, and incredible... all possibilities are possibly true! That is, all possibilities are true until a human, interacts with the photon, the glass of water, or the cat in the box. This is scientific evidence that our brain's awareness and interaction with our surrounding environment, is determining a perceptible path for us. It has been shown that human observation of natural phenomenon, actually changes the way our surrounding environment behaves (Weizmann, 1998). Humans have measured how electrons behave during the famous "Split Screen Experiment." A single electron is sent through an initial screen with two separate slits. The electron, all alone, is a jumble of possibilities, it could behave as a wave, or a particle, and its position is variable (Rogers, 2005). When observed, the electron is seen as located in the wave in a single position, observation changes the characteristic of the electron. As with a single electron, there are more likely positions for an electron to be found in different waves. A wave function is calculating the probability of seeing a particle in any given position when we look at it. Every possible outcome happens one hundred percent of the time, that is, until a human interacts with the environment, then a path is chosen. For some, this influence is clear evidence of free will. Our entanglement with our environment shapes our very existence. This "fuzzy," undetermined superposition of subatomic particles, only come to a concrete result when a human interacts with the particle (Kocabaş & Çelik, 2024; Sánchez-Cañizares, 2019). Objective reality is in "superposition," every possible position is possible. It is a possibility that our entire knowable world branches into a new universe, (constantly, automatically, and often) but the branch that we are on, is the only existence of which we are cognizant (Carroll, 2019).

To discern these ideas and other laws of physics, math equations have often preceded the actual observation in reality. Logical/mathematical thinking preceded: 1. the discovery of Earth not being at the center of the universe, 2. black holes, 3. Kepler's laws of planetary movement, 4. Maxwell's prediction of electromagnetic waves, 5. Bohr's idea of quantum wave function, and the list goes on. Current thought is that our big bang, our solar system, galaxy and universe, is probably not the only universe. The particular arrangement of energy, forces and elements that made our universe, probably occurred elsewhere and made other universes. In the realm of infinite possibilities, mathematically, there could be another universe, another galaxy, another... you (Carroll, 2019). Our Earth is at the perfect distance from the sun to allow for what we know as life. This position is one of an unlimited number of positions. Could Earth's position be an uncanny accident? Did Earth's position *have to* happen because this position is one that *could* happen? Some physicists currently think, yes, not only one other duplicate of you and your universe, but an infinite number of you, and an infinite number of universes. In a practical sense, the choices you make, the movement of your body through this existence, is the only one of which you are knowledgeable. There could be other versions of you making other movements and other decisions (Carroll, 2019). Full disclosure: I am not a physicist. However, I will reference real physicists and other scientists' work. What is attempted here is to understand the logic, synthesize scientific findings, and look at the ramifications on philosophical approaches to living. The fact is, our world is definable through mathematical/logical design.

Entanglement, is a word for ourselves interacting and influencing our environment, there is a real possibility that we create our reality in our own brain. When we exist in this earthly realm, we affect the environment around us. It's certainly true, that only our perception of this existence is true to us. This perception is understood and represented to us by our own individual brains. Focusing on what we create, as humans in our creative imagination, may very well be the template of our own possible creator(s). Just as the computer brain mimics our own human brain, our human brain mimics our creator(s). It is also true that our brain's function/perception could be a random occurrence, with no meaning whatsoever. We will explore the likelihood of both of these possibilities.

There are many possibilities for this life, it could just be imagined, a façade, or a dream. It could be real, or design, or a statistical inevitability. These are all debatable possibilities, however, we all certainly have a perceptible reality. We humans seem to have, *at the very least*, the illusion of control of our own actions, control of our thoughts and our words. I could slap someone, run away, or burn this book I am writing. It's certainly real to us humans that we could pick any course of action we desire. My senses definitely tell me I have a multitude of options, or choices of action, from which I can pick. Our perception of free agency, free will, to choose what we do, is real. Our perceived reality, our experience *is* our reality. Our subjective experience is the only way we interact with reality, and due to this fact, our subjective reality is valid, because predestination is not our actual experience. We are actively choosing and participating in what happens throughout our perceived existence. Time, in this life, only goes forward, no one (yet) can go back and change the past, but you can, in this moment, choose a path to take. You can make immediate choices to determine "which road" you will travel upon.

Our constant life choices are influenced by cultural and social norms and acted upon, or not, by considering the possible consequences. We are influenced also by our environmental situations and our beliefs. However, every single time, we have the perceptible choice to act, and/or to think in one way or another. The key here, and our academic/logicians will bristle at this statement, our emotional gut guides most of our choices (Ataei, et al., 2024). Our feelings and urgings, have massive impact on what guides us to any belief, or decision (Blackledge & Hayes, 2001). The emotional affirmation of thought and action rules our path. We are destined, façade or not, to a life of making choices that are influenced by our feelings toward anything and everything. Even the choice to believe in nothing, or to do nothing, are, in fact, choices.

There are practical (and beneficial) reasons for free-will being a reality. If there is no possibility of choosing our own path, choosing our own actions, there would be no feelings about what is deemed as right or wrong. We would have no emotional reaction to any choices we make, but we do. We would have no inner feeling of accountability, but we do. It is possible to have some loss of control in life choices. For example, if some powerful tyrant has control of our lives. We also lose control of life choices from our culturally conditioned responses, we are taught to repress threatening thoughts and/or transgression of some action we deem as sin, or to be wrong. Our own conditioned thoughts could be part of our knowledge-limiting box. In organized religions, or small cults, it is taught by leaders, the haughty-know-it-alls, that repressing thoughts, impulses and feelings, are sometimes touted as beneficial, and godly. Repressing thoughts, impulses, and feelings is viewed as a "correct," "right," or the "good," mode of operation to many. Some powerful religious leaders would say something to the effect of, "Willful obedience to (my particularly chosen) moral law is good." Regardless of the amount of mental and/or physical domination by others, we have the capacity to make any choice we mentally decide to undertake.

Chapter 4

Mind Cloud, Universal Consciousness, and External Brain Waves

Quite incredibly, when the Wright Brothers were learning how to fly in the early 1900's, people in Germany, Brazil, and Paris were working on the same thing (Garrison, 2003; Jordan, 2003; Farman, 1908). There is a dispute as to whether the Wright Brothers really were the first in flight. Similarly, when humans first harnessed fire, it's conjectured that lots of tribes in different areas of the world were harnessing fire at the same time. Why? The invention of, sewing needles, woven cloth, rope, basket weaving, boats, flutes, all these inventions show up in all parts of our globe, in cultures that had no knowledge of each other, and were *not in contact* with each other (Frater, 2007; Fagan, 2004). There is considerable distance in time between cultures that learned to work metal and use the wheel, but still, distinctly separate, unrelated cultures, from everywhere around Earth, came up with the same exact inventions (Buchanan, 2024; Gambino, 2009). This is clear evidence that there is some sort of unity in human consciousness, *between* people, even if they are not in direct communication with each other. Historians and scientists separate the ages of prehistoric eras. The eras are divided into Paleolithic/Mesolithic/Neolithic/Iron age; the divisions of eras are aligned to human technological advances. Sociologist Robert Merton, expounding on Francis Bacon's ideas, found that discovery and innovation are inevitable as knowledge accumulates over time (Merton, 1961). "The overwhelming majority of inventions… are in fact developed by individuals or groups working independently at roughly the same time." (Lemley, 2012. p.5) A collective, unseen, "mind cloud" was around (and continues) long before the invention of the internet. Since we are limited by our current space-time reality, we do not understand what the neurons of our very brain are representing in objective reality, but here is compelling evidence that we humans are connected in thought.

This evidence inevitably leads any rational person to the idea that there could truly be a "collective consciousness," where the thoughts and creations of one person, actually could have an impact on all of humanity. It is possible that all of our separate brains are picking up unseen waves/ideas/knowledge from a universal source. Our individual brains could be actually functioning as an antenna. Picking up on, and deciphering waves from a collective source. The same inventions and discoveries, occurring at the same time, lend credence to this possibility.

Throughout the late 1800's, through the mid-twentieth century, there was groundbreaking, scientific research on radioactivity and uranium. Communication at least *could* happen in this time period. Is it coincidence that at the same time: Germany, Poland, Britain, Italy, Austria, Sweden, the United States and other countries, started delving into radioactivity, and the use of uranium/nuclear fission in varying bomb apparatuses? Of course, the race was on, to use nuclear fission (Burton, 2020). Historically, no one had **any** knowledge of radioactive particle emission, yet within one hundred years, throughout the twentieth century, countries all over the world starting creating nuclear weapons. Yes, the technology advancement was reached quickest by Allied Forces, and the USA utilized its destructive power first, but still, it's another clear example of thoughts, in human minds, inexplicably "discovering" the same technological advancement at the same time.

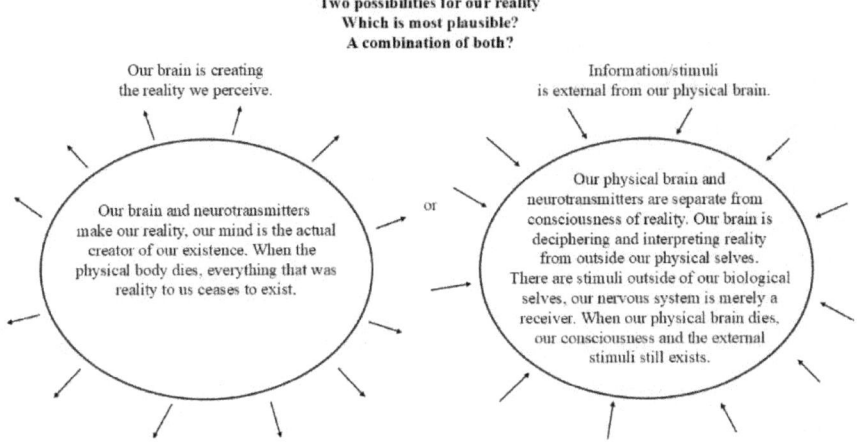

The circle diagram shows two possibilities. One could hypothesize a reality where there is a blend of these two possibilities. Our perceived environment is either 1. perceived from an external, true reality, or 2. our perception is created from within our brain. Our brain is the integral part of our consciousness. The brain has been proven, in some cases, to create conscious experience for us humans. Electrical current has been passed by an exposed part of the brain, the awake person sees a "lightning bolt," and even shapes of letters pass across their field of vision (Beauchamp, et al., 2020). Electrical current to the human brain can also give a person the sense they are truly, physically touching something, when they are not (Bowling & Banissey, 2017). In these cases, the biological (brain) comprehension (or creation) of an experience is perceived as something separate, but this is proof that our brain is creating our reality. People with missing limbs have itchiness and go to scratch what is not there. A reflection inside a mirror is an extension, made up by our brain. The ability of our senses and nervous system as interpreted by our brain is limited. This is why we use other instruments to see that which is normally undetectable to the eye. Items such as radio waves, microwaves, infrared, ultraviolet, x-ray, gamma, none of these are perceived through our un-aided eyes. Yet, we can see these waves with the use of created instruments to help our deficient eyes make a visible representation of what is actually occurring. Our sense perception is the best tool, it is the only tool, we have to discern reality. We rely on our sense perception because this *is* our reality. Part of the mystery of why we exist is linked to this basic self-inspection. Just knowing the word "reason," thinking, pondering, and deciding, we are conscious of doing these thought processes. Our sensory intake (or sense creation within the brain) is our reality, we are forced to use the only tools we have.

There is also evidence supporting consciousness being something separate from our physical brains. Consciousness being something outside of our physical selves. The aforementioned "mind cloud" is one example. Consciousness may be something we are tapping into. In addition to the mind cloud, there is proof of external consciousness merely by our existing through constant life experiences. We are aware of something outside of ourselves. We are experiencing things not created in our brain, but rather, interpreted by our brain. The proof for this is, we humans have an objective, shared, waking experience. This is easily seen in our language, in the words we use, the words mean something to someone else. We can say the word "unicorn," and everyone knows what that is, even though it does not exist in reality. Outside of our own brain, others will know what is meant when "unicorn" is said. Even in other languages than our own tongue, every language has words to represent similar objects and ideas in their native tongue. This can only happen because we are sharing a similar reality, a separate reality. A separate *thought realm*, a consciousness not only limited to our own physical neurons, firing in our own brains, but we are in a shared world. A word stands for something in our experience, say, the taste of garlic, or a saxophone sound, or a color, the words point to a "thing" we experience (Duqum, 2024). The word is nothing by itself, yet the color, or garlic, or sax sound, or whatever, is an experienced feeling, our history of interactions are attached to the words. What if we told another person the word, "oregano," but that person never had a single experience with oregano. They would not know the reality of oregano: what the plant looks like, what color it is, how it tastes. We only know things of which we have some interaction and understanding (Fesce, 2023). In addition to having this shared world, we do not even see this world objectively. We do not interact with the subatomic reality that is truly running *everything*, but the invisible-to-our-eyes subatomic reality, is certainly present behind our limited brain interface. We do not experience anything in this life as it truly is in our shared objective reality (Duqum, 2024). If consciousness is external to our physical body, our continuing, or ceasing to exist, when we die, is not definitively provable either way. If our consciousness is external to our physical body, when we die, it is possible that our consciousness could go on in some entirely different mode. This line of reasoning does equate our consciousness to a possible soul/spirit that is separate from our physical selves.

Our consciousness, as we exist in this life, has greatly varying modes of operation. Awake, asleep, amnesia, anesthesia, coma, brain trauma, psychedelic drugs, in all of these states of consciousness we are still alive, yet existing in completely different ways (Murray & Srinivasa-Desikan, 2022). Measurement of brain activity (electroencephalography, EEG gamma coherence) shows connectivity between different regions of the brain, neurological variances in "frontal coherence" for mind/consciousness is clearly seen when people are in deep meditation (Ken et al., 2022). In transcendental meditation people feel absence of time, space, and sense of their own body (Travis, 2014). This deep meditative, measured, brain activity is different from our waking conscious, which is also different from our sleeping/dreaming conscious. Dream consciousness is solo, everything that happens in our conscious dream state is up to us, individually. Conversely, during waking conscious time, our conscious minds are interacting with other people who are in an awake state. Another difference between awake and asleep conscious times are the changes in the natural laws of physics. We are subject to physical limitations when we are awake, some of these limitations do not exist when we are in our dream conscious state. In our dream conscious time, we are not beholden to limitations of movement (we can fly, or be stuck), and time can be distorted. One state of being (awake or asleep) is not *more real* than the other state, we are alive during both times, but we exist in different ways. In both waking life and in dream life, we are still surprised, we are still emotional, there are chaotic moments, and we still make choices. What makes us, us, is still existing whether our consciousness is functioning while we are awake, asleep, meditating, or in some other altered conscious state.

Further scrutinizing the levels, or modes, of consciousness while we are alive, let's dwell on the times we are anesthetized, or in a coma, or when we are physically knocked out. It is questionable how we exist in these moments. During anesthesia we have very limited, or no state of consciousness. These moments when we are under anesthesia, we are still alive, still existing here on Earth, but there seems to be no time for the person under anesthesia. We are not aware of our self, we feel nothing, no pain, or pleasure, no emotion, and we have no memories during these "brain-inactive" times when we are anesthetized. So, as when we are dead, under anesthesia, or in a coma, we are not creating or interacting with our environment. We have no knowledge of our own existence, yet, we are still existing. Something of us is existing in the minds of other people that have knowledge of our existence, they can still see our physical body. Physicist Roger Penrose and anesthesiologist Stuart Hameroff, have postulated that there are multiple possibilities in microtubules (a part in the cell) orchestrating our conscious experience to a single definite state (Hameroff, 2021). Understanding, knowing, being alive, comes through *our* consciousness. If our brain is an antenna picking up consciousness from somewhere outside of ourselves, then the waves are not being received (as they normally are) when we are in a coma, or anesthetized. This is easily observed when compared to the moments during our times when we are in waking consciousness. If consciousness is made from our very own brains, when under an anesthetic, our brain is not creating, or receiving information. If consciousness is purely a creation of our own brain, it is an interesting dynamic that other brains can still experience the anesthetized person as a living person. Brains have knowledge and shared experience with the other thought-filled brains of our world.

It is debatable as to whether waking-brain consciousness can be born out of the human creation of intelligent computers. Can a computer programmer create a computer that is able to have similar experiences we humans have? Similar to what we humans experience while existing here, in this life? Some say "yes," some say "no." The "yes" people would point to the fact that our biological brains can be electrically stimulated, this electronic stimulation creates a vision in the person's perceptible field of vision (Beauchamp, et al., 2020). If neurons can do this, electrical impulses in computers may be able to do this as well. The "no" people would say, all your experiences, your neurons inside your brain creating these interactions, are not even real, but a dumbed down version of something infinitely bigger, with which your brain is interfacing (Hoffman, 2019; Shermer, 2015). Circuits and computer software do not exist without human perception and creation. If our human brain (neurological/biological) doesn't create consciousness, computer programs can't either. All our perceptions in our space-time realm are only representations made by us when we observe anything. Every subatomic particle we measure has no value, until we humans measure it, the particle is in *our* context, our projection, our construction

(Duqum, 2024). The billions of neurons that make up our brain create a scaled back representation of an unfathomable consciousness in reality (George, 2020). Therefore, it is possible that death could be the shedding of our limiting brain (and body) and going to the infinite consciousness that is only housed in, and represented by, our physical selves. If consciousness is separate from our physical self, a version of our conscious self could continue, but be very different from how it currently exists and functions while we are in this life. A seed, sprouting through the dirt and growing into a plant, could be a crude parallel to our earthly conscious growing into consciousness after our physical death. Earthly consciousness… to… next existence consciousness. A tadpole to a frog, a caterpillar to a butterfly, these are possible, visible, physical parallels to our earth conscious changing into next-existence consciousness. We could *change* in the way we exist. These are all just reasonable physical comparisons to how it could work. The invisible realm of conscious-self, continuing in a different capacity than the way we understand it in this realm.

There are other evidences of consciousness not emanating from only inside our own, physical, human brain. Electroencephalogram (EEG) is a human made machine to see brain activity. The amazing fact is; these scientific EEG machines have electrodes that are placed outside the brain. Electromagnetic fields from the brain can be read by computers from **outside** a person's head, and deciphered! Brain-Computer Interface (BCI) systems are used by people to operate medical devices, and to aid in breathing (Golee et al., 2010). There is another recent invention, a car that runs only from an individual human's thoughts (Yu et al., 2016; Wang et al., 2019). Electrical signals are read, that are emanating from the human brain. Humans can already convert these electric brain signals into commands that can control and move cars, arms, legs, and even to communicate with words from a person who does not have the ability to speak (Hamzelou, 2022). This breakthrough in technology is using electrical waves *outside* the head to accomplish these tasks. This is clear proof of brain waves/consciousness outside our physical selves.

Scopaesthesia is another instance where the brain has unseen activities separated, away from its physical self. Scopaesthesia is when you feel someone is staring at you, you know it without your eyes seeing it. A person will instinctively feel someone's stare, and turn toward the person who is looking. This happens to almost all people. Amazingly, in scientific study, 68% of the time, the person being looked at attends in the exact direction from which the viewer is looking. We can "feel" the direction from which someone is staring at us (Sheldrake & Smart, 2023). This is further evidence that this brain of ours is picking up on something outside of itself.

In the past few paragraphs, we have talked about our brain and whether it is perceiving, and/or creating our world. We covered if our consciousness is from an external source, or a creation of our brain. We deliberated the possibility of a collective, universal consciousness which we all may possibly be "tapping into." All of these topics have direct implications on what happens to our conscious selves when our physical self perishes. The idea of our very selves, what it means to be an individual, a conscious person, is paradoxically both a mysterious question and, simultaneously, our obvious reality. Our personality, our preferences, the psychological aspects of us as individuals, is a part of our consciousness. The bigger idea of us, as an individual essence carrying on after our physical lives here on earth (or before our perceived time here on Earth) might be understood by analyzing these iterations of consciousness while we are alive, here on Earth.

Chapter 5

Time & Space

Current scientific study of physics and delving into the nature of time, may shine some light as to how we perceive existence. Albert Einstein's' theories, some already proven, some disproven, state that time doesn't flow in one direction like we think it does. This perception of time is an illusion given by our physical selves being limited to a particular place in our space-time continuum. Any constant and regular repetition, of any physical object, can be used to measure the passage of what we call time. For example, the Earth spinning is one day, orbit around the sun is one year, the phases of the moon, can all be measured by whatever we choose. The weird thing is, this idea of "time" is different for every individual. When a person, or planet, or whatever, is moving, this movement *slows down* the passage of time for them (Greene, 2004). Time is not a constant truth; time is a creation of our brain. When humans go under anesthesia, the electrical impulses between neurons are blocked. There are no memories being made under anesthesia (as well as people in a coma), the passage of time does not occur to the person with the receptors in their brain blocked. When a person under anesthesia comes to conscious waking time, it is as if no time passed at all. Each person, creature, plant, planet, star, rocks... all things, have their "own time" dependent on their motion through space and the receptors in the brain making sense of the world within this framework of time and space (D'Ariano, 2017).

One moment in time, in one place in the universe, is a different moment from another thing/sentient being in a different part of the universe. There may not ever be a true "now moment" for anyone as we perceive it, but rather, multiple moments are occurring simultaneously (Greene, 2004). Motion and gravity, our position hurtling through space on this ball we call planet Earth, affects our individual perception of the passage of time. So, when we look into space, at the stars in the night sky, the light that is hitting our eyes has traveled many light years to get to us. Light travels at 299,792,458 meters per second (1080 million km/hour; 671 million mph), that star light is an event from the distant past, the light created by the star already happened. In this case, when we see very old star light, in position of space time, we are actually far, far ahead in the future to when that starlight was created... it's just now being perceived by our eyes. Occurrences happening to our perception is an individual viewpoint. All the occurrences that have ever been, or ever will be, already exist, all at once. This has huge ramifications when we discuss our death. When we die, the reality is, we still exist, depending on the position in the universe of some other perceptive being. All of our past, and future actions, still exist, right now (Williams, 2017).

My son Daniel, is a computer engineer. He was telling me, interestingly, a common way to glitch videogames, is to move an object very quickly. Since everything in a game is on an x,y,z grid, when you move an object along a straight line you get the classic $y=mx+b$ behavior in a time slice, or time frame. Let's imagine we have a plane (a wall) at position "10,0,0" that is one unit thick (from 10-11 on the X axis) and you move a point along the X axis, starting at the origin. If the point is moving one unit per frame, it goes from $x=0,1,2,3...10$ where it collides with the wall and the game's created physics stops it. Now, let's move the point slightly faster at three units per frame; it goes from x=0,3,6,9,12,15... Look at what happens by moving at intervals of three. We did not collide with the wall at x=10-11! We just "glitched through" a brick wall within the video game! Game engineers have to account for this in various ways, for example: programming "stop if x>10." The fact is, within video simulations these unexpected loopholes exist. While not a true comparison for quantum tunneling, it is analogous to the glitchy behavior that could break seemingly constant rules, and allow for humans to sidestep our time/space confinement.

Time during our life, in a more practical sense, the way we actually experience time, resembles a flowing river, it just keeps rolling on in one direction. When we are young, who cares about time passing? No child cares about time. However, these life moments become precious beyond belief as we get older. How many hours have I slept? Played games? I remember devoting full days and into the night, playing cards. I was addicted for a long-time to playing chess. How many hours have I sat looking at a computer playing chess? How long have I sat banging on my drums? Work, there are some days at my work where I talk with young people and I feel like I had some important impact on a person's life. Other days, I sit in meetings that are completely useless. As we age, an urgency starts to enter our lives. Time is such a valuable, yet, oddly aloof aspect of this life. Later in this book, we'll delve into the subject of whether there is a hierarchy of actions, when using up this perceived, valuable, entity that we call "time."

The time and space construct by our brain, allows us to live in this complex world, which is really made up of quickly moving subatomic particles. Another video game analogy is helpful, this idea is from Donald Hoffman (Hoffman, 1998). When we humans play a video game, we do not see the real coding, electrical exchanges, we don't see the inner workings of what is really happening inside the video game.

Pondering Our Existence for Practical Living

We have a "dumbed down" headset, or hand controller that allows us to live and move within the video game, efficiently. If we had to know all the inner workings of what is happening, we would be stuck, we would not be able to function. All of the toggling of millions of voltage variances, we would not function inside the reality of what is truly happening inside the video game. In the same way, our brain is analogous to the hand controller of a video game, the stimuli through our senses are "dumbed down" for us to function in this space-time realm here on earth (Hoffman, 2019).

A metaphor concerning our brains' limitation is instructive: when we flush our waste away in the toilet. We know how to function, (as our brain knows how to function) but we may know nothing of what is really occurring. The float in the tank, the neck twist in the toilet bowl, the level to shut off incoming water, the pipes out to a septic system, the drainage system through pumps, the chemicals to treat the sewage, and then the return drainage back into a river. We may not know all the "magic" behind the plumbing, we just know how to function by flushing our waste away down the toilet. Space-time is a façade for what is truly occurring beyond our limited neural and sensory abilities. Space-time is an illusion made by our own brains so we can just function, as the toilet flush just functions. Einstein's theory of relativity for space-time mathematically does not even exist when you divide space to the tiny 10 to the -33 centimeters. Also, when you divide time to a much-smaller-than-a-second (to 10 to -43 seconds), space ceases to exist (Arkani-Hamed, 2012; Kuhn, 2021). We know this mathematically, we do not see reality as it is, space time is **not** fundamental. All objects in this space-time realm in which we live (clouds, trees, stars, air, birds, tables, your body… *everything*) do not exist in the forms we know. Everything is perceived by us through our brain, but all things are functionally different in objective reality. They are really electrons, protons and neutrons (and smaller subatomic particles) all zooming around. When all things are not being perceived by us, their state of existence is extremely different. The same truth happens during the quantum (previously mentioned) split screen experiment. The photon acts differently when a human observes it in our space-time realm.

Since time and space actually "collapse," they cease to exist, when minutely divided (Arkani-Hamed, 2012; Kuhn, 2021), there is likely to be a true reality, separate from our space-time limitation. There has been discovered a mathematical, jewel shaped, geometric structure that suggests there is reality outside of our space-time existence. Scientists have discovered this, after technological advancements in the ability to take a single electron and collide it into a proton. When tiny subatomic particles are purposefully smashed, the electron scatters off and scientists measure where it lands. This is called a "scattering amplitude." The tiny subatomic particles quarks/leptons and gluons/bosons are inside protons, these interact to balance the state of the protons and neutrons (D'Ariano, 2017). When scientists intentionally collide electrons and protons the electron will exchange a photon with a piece of the proton (either a quark or a gluon) the quark scatters in a three-dimensional way, then scientists measure where it lands. When they first started doing this, scientists plotted (on a histogram) where these subatomic particles landed. It took hundreds of pages of calculations to mathematically come up with a prediction of where the particles would land. Beyond any space time calculations, scientists found that a single, geometric structure, an obelisk with edges (now named an amplituhedron) now accurately predicts the scattering amplitude that once needed hundreds of pages of calculations (Franco et al., 2015). This jewel shaped geometric structure, the amplituhedron, is evidence that there are logical/mathematical structures beyond that of our space-time reality (Hoffman, 2024; Arkani-Hamed, 2019).

Chapter 6

Design, Chaos and Infinite Possibilities

In William Poundstone's book, *The Recursive Universe,* Poundstone puts forth the idea that in an infinite setting, everything that is possible **must** happen at some time, in some place (Poundstone, 2013). A similar idea plays out when we look at the ratio of the circumference of a circle to its diameter, Pi (π). Pi is an irrational number, it has evident randomness, yet, illustrates that all possible patterns of digits occur somewhere in the expansion. Pi is enlightening, any finite sequence of digits you could think of, will be found in π. For example, a single number repeats, or an order of numbers repeats, π will have sets of numbers that repeat and have order. Back in the 19th century, John Venn decided to show an example of randomness by using π. He assigned cardinal directions numbered 0 through 7 (Venn, 1962, p115). My son, Daniel Gentile, and I, decided on a similar experiment with π. We assigned coordinate directions to each digit (0,1,2,3,4,5,6,7,8,9) in π: 0 was straight "north" a 90-degree angle. The digit 1 went NE at 36 degrees, the digit 2 went NE at 72 degrees, the digit 3 at 108 degrees, and so on…

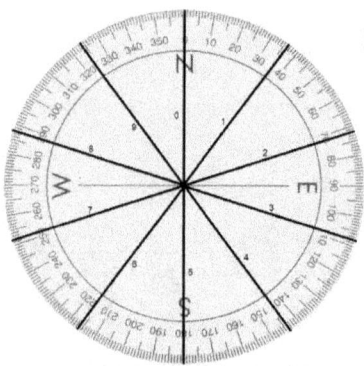

We started at the origin of an x/y grid. The design of assigning cardinal direction to the digits of π points this out nicely. The following graphs are what comes out with each term of π being assigned a cardinal direction.

Here is a graph when cardinal directions are assigned to the first one hundred numbers of π.

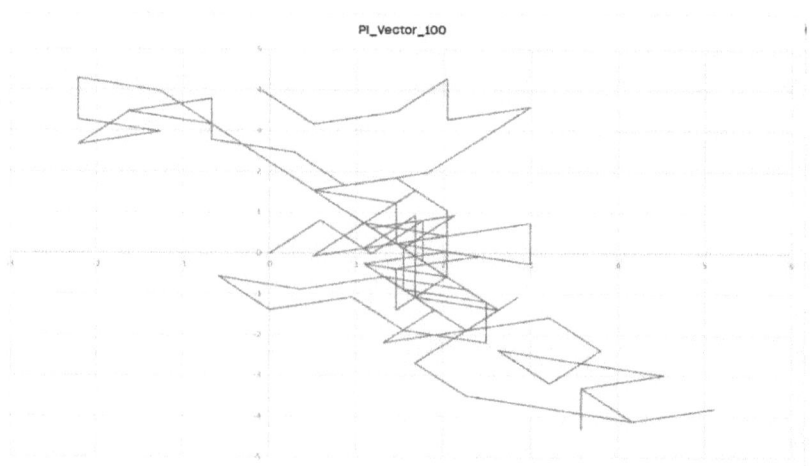

Some of the areas in the above graph look like an illusion painting, triangles, straight lines, but it's mostly a jumble.

Pondering Our Existence for Practical Living

This graph includes the first 500 digits of π:

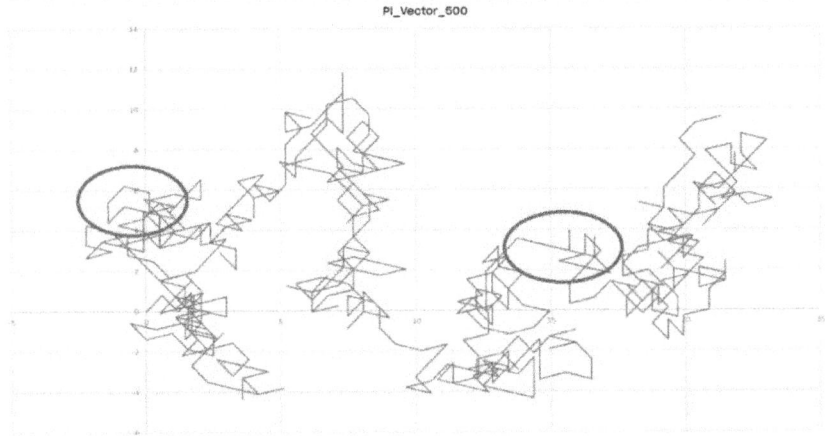

It looks chaotic for much of the directional representation, the circled areas (and others) have straight line direction and perceptible geometric order.

This graph includes the first 1,000 digits of π:

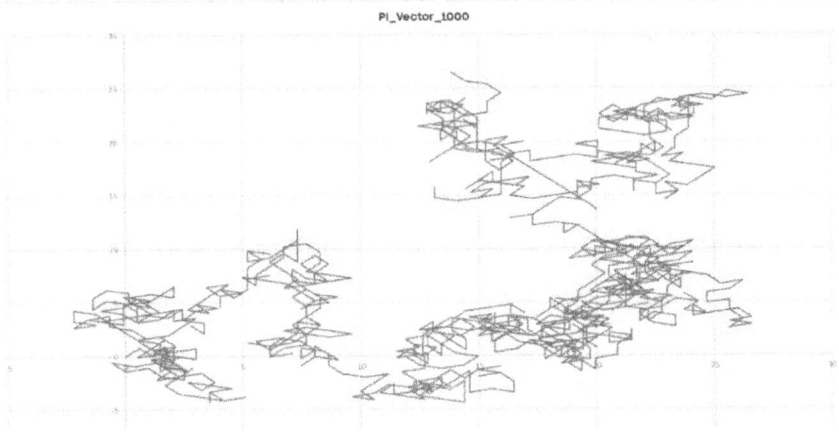

This graph includes the first 10,000 digits of π:

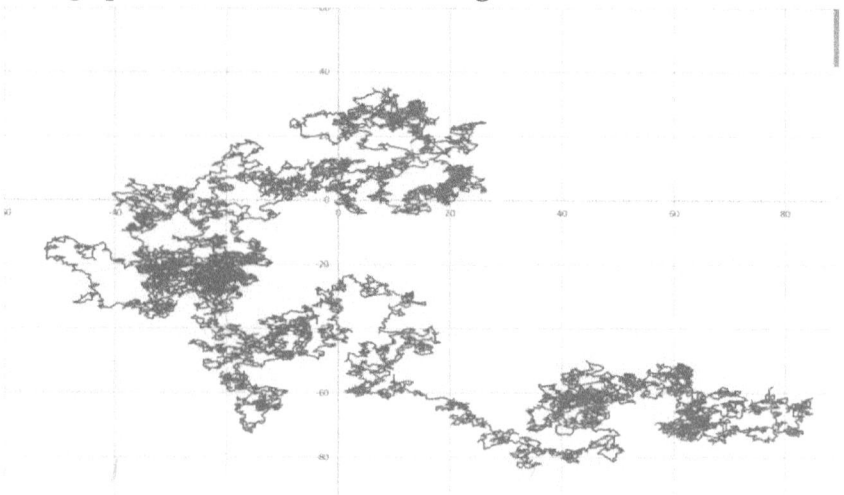

The design that came out from assigning cardinal directions to the numbers of π was mostly chaotic. However, in certain, areas, true design would show up. Giving cardinal direction to each digit for the first 10,000 digits of π, overall, creates the image of a meandering trail, hooks, left, goes right, goes downward, etc. Either some creative entity, or just infinite randomness, created this realm in which we live, and this realm is just functioning according to its own laws of physics. According to Poundstone, in any infinite plane, every possible pattern must eventually occur. In other words, if the possibility of all parameters can occur to create life, then somewhere, life must occur. The random chaos of π is evident, during certain sections in the expansion, π does have perceptible design show up. Chaos and order, just like what we discern in our own world.

There are many incidents of chaos when compared to the areas of order in our universe, yes, even in our very existence. Take an electrical extension cord, or a necklace, or garden hoses for example. They all become entangled at the slightest piling of them. Why can't we pull on a cord which we have just piled together, and it come out in a straight line? It always knots and tangles. Why does everything rot, rust and decay? Another example, a piece of paper. Let's rip a normal, letter size piece of paper into small pieces and throw the pieces in the air. What are the chances of the pieces of paper coming down through the air and re-assembling itself, in the exact configuration of pre-torn paper? A person could take up the ripped pieces of paper and throw them into the air over and over again, an infinite number of times. How many times would those pieces re-assemble in the same order before they were torn? This is a more realistic explanation to understand our existence that puts doubt in the everything-must-happen-in-the-realm-of-infinite-possibilities explanation. The random-must-happen-because-it-could-happen theory is shown to be a remote possibility. The re-thrown, torn up paper, eventually falling back together, (miraculously) into the previous ordered/un-ripped paper, may actually be impossible. It seems probable that the paper would **never** fall back into that exact order. In the world of infinity, it is possible some occurrences would never happen.

As shown above, there are difficulties in believing that chance, and infinity, are the reason behind our own knowledge of our own existence, this just seems quite implausible. How can infinity give us the self-knowledge that we exist? How can infinity allow us discernment of chaos from order? Are we to believe life itself randomly sprang from chaos because it had to over the course of infinite possibilities? Even if that is true, it does not explain what most current physicists believe today. That is, that the Big Bang created everything. If at the start, at the moment of the Big Bang, there had to exist what we conceptualize currently, scientifically, as strong gravity that changes the structure of time and space. There were subatomic forces, temperature, particles existing we call bosons and fermions… everything. Does there *need* to be a creative entity for these forces and items to even exist? Can these forces come from nothing? How and why did these ingredients even come to be present at the moment of the big bang? Current thought of leading physicists is that our universe is expanding at an accelerating rate. Physicists think there is something called "dark energy," aptly named because it is unseen by any of our instruments which enhance our own vision at this time. This dark energy is pushing galaxies apart ever since the big bang, which ostensibly occurred about 14 billion years ago (Goldsmith, 2008; Hadhazy, 2017; Mitton, 2006). Similar to our current inability to "see" dark energy, humans do not know for sure what is happening at the microscopic, subatomic level. Inside atoms are protons and neutrons, which are, in turn, made up of "quarks". Some scientists believe quarks are made of something even smaller, tiny, vibrating strands of energy, scientists call "strings," voila - String Theory (Antoniadis, 2015). These incredibly tiny strings vibrate, could this be why music touches something, emotionally, within us? Anyway, the theory is that these strings make up our entire universe, not just three-dimensional length, width, height, but other dimensions. Explanations of objective reality are continually being pursued in the realm of physics. To deny the obvious design/chaos interplay that is used by everything, every day, and studied by scientists every day, one would have to deny studiable reality.

Chapter 7

Science & Logic Guide Our Understanding, Emotion is Powerful
A word about music

Music is mathematical, any rhythm is dividing our passing of time. Music is this, plus our brains' interpretation of pitches at different wavelengths. Even though music is scientific, music also moves the heart, it moves us emotionally. Why is this so? Since emotions are a powerful guide to our every choice in decision making, it would be wise to spend just a minute to question… Why does listening to music make us feel anything? Just music notes, even without words, causes us to see images, or to have emotive feelings and thoughts. Here are some ideas, borrowing from ideas of the vibrations in String Theory. In String theory, subatomic particles such as electrons and quarks are vibrations in quantum fields (Bicudo et al., 2022). The vibrations at the subatomic level could have an effect on our emotional state, in the same way, vibrations of music waves through the air give us emotions. Our world can be understood by mathematical/logical formulas. Music, at its base level (vibrations in wavelength, plus rhythm) *is* math, thus music is illustrating to us, in an obvious, and easily understood format, the very fabric of our existence. This is why music is able to resonate within, and affect our emotional state. On an observable, scientific level, music, neurologically lights up the brain in magnificent ways (Parsons, 2005).

The distance in sound wave frequency creates the musical note's pitch that we hear. Think of the look of surface water when you throw a rock in, the ripples go away from where the stone entered the water. This is the same way sound waves move through the air. We can measure the tops of those "ripples," the waves of sound moving through air, this is called the wavelength. High pitched sounds, lots of waves per second, or hertz, make us feel emotionally different than low sounds with few waves per second. The same goes for the tempo of music, or the speed at which notes are played. All of us are subject to time in this realm. The distance in time between notes dividing the time in which we live, will give us different emotional responses. Even if the notes do not have tone. A drumroll with quick successive strikes/hits on a drum, the emotional energy rises, there is anticipation. On the other hand, if a drum is hit slowly, with longer spaces of time between hits, the feeling becomes mellow or menacing. This is in direct contrast to the hyped up feeling of quick notes. Basically, when we feel these differences, it is due to the fact that the musician *is dividing the passing time* in slow, long spaces, or quick little spaces. We are dividing the "now" moment of our existence. Music influences our emotions and our experience of time, it cannot be ignored (Burger & Wöllner, 2023).

The importance of our emotions, our feelings, are often maligned in this day and age. "Hide your feelings," is advice often given. You want to win arguments? Make deals? Get along? Well, then hide your emotions. It is pervasive these days that decisions should always be based in fact, in logic, not emotion. The reality is, all decisions are highly impacted by our emotional reaction. Our emotional reactions scream loudly within ourselves as to what we feel, and when something feels right or wrong. Contemplation of what is beneficial, or what is harmful, worry over decisions, deliberation over our actions, is derived from our individual judgment of what is right and wrong. What is good and beneficial, or bad and detrimental, is ever-present in our psyche. Since emotion is always present, this aspect of ourselves should be recognized as impactful as science and rational thought. Since music is tied to emotion, this is also why music is so important, it prompts an emotional state within us.

Aside from the emotional impact of music, and emotional impact on our decision making, there are biological benefits to music as well. The corpus callosum is a large bundle of nerves connecting the two hemispheres of the brain. Musicians have a significantly larger anterior section of the corpus collosum when compared to non-musicians (Burriss & Strickland, 2001). The larger size allows greater communication between frontal lobes (Strickland, 2002). These neurological findings lead to discussion of brain laterality, which is how the hemispheres of the brain operate. The brain is quite literally changed by musical experience. Myelination, is when a fatty sheath increases on neurons which allows electrical impulses to transmit through our nervous system. Musicians have greater sheath density, and an enlargement of motor and auditory related brain structures. Musical training actually changes the structure of our nervous system (Zattore & McGill, 2005; Scripp et al., 2003; Skoe & Kraus 2012). These music related brain circuits give us biologically, what is felt as rewarding (dopamine) stimuli, as well as commonality with survival-related systems (Zattore &McGill, 2005; Gabrielson, 2013).

Chapter 8

Design, Equilibrium, Chaos and Order

Simple equations create complex outcomes. Complex structure arises from the use of simple mathematical rules. Using equations to plot a state where opposing influences are in balance, an equilibrium, or homeostasis, the Mandelbrot Set is another indicator of how our realm functions. As rates change in the Mandelbrot Set equation, equilibrium splits in two. There become two outcomes as different values are continually fed into the equation, the outcomes split again, into four outcomes. This pattern continues, the outcomes keep doubling, these "doublings" are called bifurcations (Muller, 2020). Eventually, just as in our discussion of π, artistic, ordered designs arise from the Mandelbrot Set. Then there is a return to random, chaotic structures. As values are continually plugged in by computer use, ordered design, amazingly, comes in and out of chaos. Lots of chaos, with little areas of order (Muller, 2020). The same with the earlier discussion of irrational π. The idea is, there are undeniable, mathematical, logical equations for how we are existing. The universe is vastly, predominantly chaotic, then, there are instances of perceivable, what we sometimes call "beautiful," design.

Pondering Our Existence for Practical Living

Physics/mathematical calculations explain our universe; these equations are proof of a design. One could believe this design emerged randomly, accidently. It's equally legitimate to believe there is some sort of creative entity. Some say this world in which we exist could possibly be an illusion created by our own minds, but either way, this is certain, it is at the very least, a complex "illusion" that we can study.

Something in our brain recognizes chaos and order. The fact that we humans are able to decide what is order and what is chaos, is evidence that there is some designing entity to our perceived existence. Our very understanding of what the word "design," and its meaning, is proof in and of itself. My son Daniel made this electronic board with glowing lights. It's apparently a popular creation for engineers to make, it's called, "Conway's Game of life." There are just three simple rules these lights are programmed to follow:

-An unlit bulb, or "dead," cell with three "live" neighbors (lit cells) becomes live. (Represents birth)
-A live cell (lit bulb) with zero or one neighbors dies of isolation; a live cell with four or more (lit bulbs) neighbors die of overcrowding. (Represents death)
-A live cell (lit bulb) with two or three neighbors remains alive. (Represents survival)

Here are three pictures from my son's light-board creation of Conway's Game of Life. Notice that with three simple rules, chaos predominates, yet areas of intricate design also emerge.

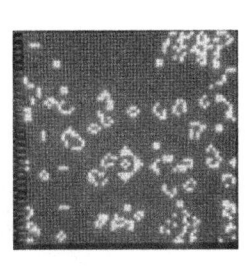 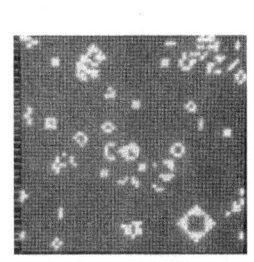 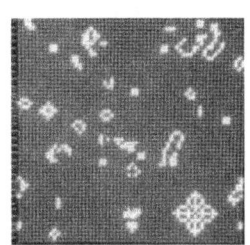

The intricate designs emerging from these three rules are discernable and sometimes beautiful. How many scientific/physics laws are perpetually running in *our* universe? These mathematical/logical laws are certainly here, they are true, scientists measure them, and we naturally use them at every moment of our physical existence. One must question; how can these mathematical/scientific laws exist without some kind of creative entity making them in the first place? They came from nothing? In the realm of physics there are few areas of time and space that are orderly. Entropy, disorderly randomness of a configuration, is always present. Chaos and randomness are the predominant state in our focus on Mandelbrot Set, π, Conway's Game of Life, and the deterioration of everything in this world, however, order *is* observable, it is here! These pockets of order, namely where we humans currently live, order is easy to see. Scientists are able to, quite accurately, predict the positions of the contents of our solar system. We know when eclipses will occur, tides, phases of the moon, we know when electrons will jump, our deterioration (second law of thermodynamics) and multitudes of natural laws are obvious and ubiquitous. Did all of these natural, measurable processes come by way of chance? Did this design come into existence because there are infinite possibilities that must somewhere occur? Reason loudly screams there is just as likely to be a creative entity to design with these perceptible, mathematical/logical formulas that illustrate the functioning of our universe. Plants, planets, stars, animals, life cycles, seasons, yes, it's obvious there is a design. Nature behaves in a regular manner, it is measurable, it is constant, it is pervasive and inescapable. The natural world is emphatically predictable and measurable. This happened by a chance fluke? Even knowing "order" in creation itself is an obvious sign that there is a high probability that this realm was set-up-by-some-highly-intelligent-entity; some kind of creative entity probably exists. It seems likely that some entity, some being(s) made this existence in which we perceive stimuli. Now, it is debatable as to how much this creative entity cares about our existence. We humans **are** "Conway's-Game-of-Life light board," our "light board," the one in which we live, has certainly been set into motion.

There is a well-known biological experiment on the Kaibab Plateau which is located just north of the Grand Canyon. The predators of deer were annihilated. Then, predictably, the deer population exploded because they had no predators. The experiment went on for over 30 years in the early 20th century. The deer ate themselves out of food sources. Eventually, there was a major starvation that reduced the deer population, and the number of deer went back to a level where the land could sustain these animals (Muskopf, 2023; Report, 1924). Even without any predators the deer population leveled off to homeostasis. The important point is, an equilibrium is always the direction of our world.

At the atomic level, neutral atoms are balanced between positively charged protons and negatively charged electrons. Electrons are kept in orbit around the nucleus of an atom because there is an electromagnetic field of attraction between positive protons and negative electrons. "Unstable" atoms will lose neutrons and protons in *its attempt to become "stable."* The question is - why? Just naturally, by observable design… atoms "try" to become stable, to be level, to equalize. *This is the way all aspects of our world operate.* There exists an observable leveling, akin to the interconnective yin and yang in all areas of existence. The idea of homeostasis, comes from the mathematical fact that we see this in nature and our universe. This is a scientifically observable fact. Our scientific understanding of movement of our solar system, our perception of atomic structure, homeostasis in animal populations, these all play out similarly in our everyday lives. Even in our very thoughts and actions. Death in relation to life, ridicule in relation to compliment, aggression in relation to acts of kindness… these are all interconnected and inseparable, and they all "try" to become stable, the same way the electron at the molecular level tries to stabilize. The same way wildlife populations balance, everything is going toward a level balance. If there exists a hierarchy of life virtues, or actions a living being must accomplish, this idea of "why" it exists is critical. This "why" is due to whatever creative entity is behind this existence has embedded balance into the fabric of our universe. Yes, this is extrapolation of known science, being applied to other areas of our existence, however, it's easily observable. Homeostasis will be sought after in all settings of our existence here on Earth. (Muller, 2019).

Chapter 9

Nihilism v Purpose Driven Design Favors Purpose

Some people seem to be outwardly apathetic about everything, very little in their life merits yearning or effort. This is the personality type that doesn't seem to care about anything, they live a very basic, auto-pilot/follow-what-everyone-else-is-doing type of life approach. These types of people will never delve into anything deeper than the basic, what I will call, the tier-one goals of life. This lackadaisical approach to life is utilized by many. The reasons for this apathy are either for: 1. being too busy, living in survival mode. 2. just not wanting to ponder the big questions. 3. pressure to please people that culturally surround this person. 4. not having the intellectual ability.

If we want valid answers, to not just blindly accept and repeat what other people told us our purpose is, or should be, it is important to listen to all ideas. Then decide on our own, to arrive to an idea about what constitutes a great life. Is our purpose as a human to fulfill the priorities of the culture in which we are born? Is our purpose to blindly accept our surrounding family/cultures' ideas? This seems to be the most powerful and prevalent path people pick for their lives, yet, objectively, this life path is clearly random and changeable. It's easy to pick your surrounding cultures' path, and there is extreme pressure to assume other people's reason for your existence. Most people choose to just take on what their surrounding culture (religion, government, friends, family) tells them their purpose is for existing. The cultural priorities of any geographic setting and time period have widely varied, so, objectively, this is not an effective way to pick your reason for life. It may be the comfortable path, but it may not be the fulfilling, or the most fruitful path.

We humans do have the ability to create the justification of our existence in our minds. Also, belief in no god, or gods, or purpose for our existence, is always just a belief. People with no belief, have paradoxically decided on a belief... of no belief in anything. We truly can't escape. We can't help but create meaning to life because it's created by your actions in each passing moment. If the meaning of your life is, "I declare, there is not meaning to this life." Well, then you just made the meaning of your life nothing. To make sense out of what we encounter, this is built into our very human fabric. If a person looks at clouds, they automatically start to see pictures. Humans cannot help making meaning, we are designed to see design. This is convincing evidence of a creative entity, we recognize, predictable order, and we are able to contrast it to unpredictable chaos. This is not to say that the chaos, or the "non-discernable-design" areas of our existence, are a part of the design as well. The unpredictable, random chaos areas are certainly a part of the overall design, but are categorized as "disordered" to the interpretation of our mind. The "chaos parts" are just incomprehensible to us, similar to the chaos being an integral part of a designed snow globe. The chaotic snow swirling in a shaken snow globe *is* a part of its design. The fact is, science continually studies the interplay of chaos/design of our realm. As in most of the points in this book, there's a precedence in nature for this reality.

Pondering Our Existence for Practical Living

Humans discern and we create meaning; this is also evident in scientific cymatics experiments. In cymatics, particles on a metal plate are vibrated with sound. We humans discern, or make patterns with our brains, the same way vibrations make patterns in nature, as we see in the study of cymatics. In cymatics experiments, a sound wave is projected toward sand (or salt) on a metal surface. When the metal surface vibrates according to the sound wave frequency, incredibly, beautiful geometric patterns occur. Different vibrations make different geometric patterns. The sand settles where the metal plate is not vibrating. The vibration makes discernable design, this is a parallel to our brain discerning design out of what it is taking in from our surrounding environment (Yun, 2024 & Murray, 2013).

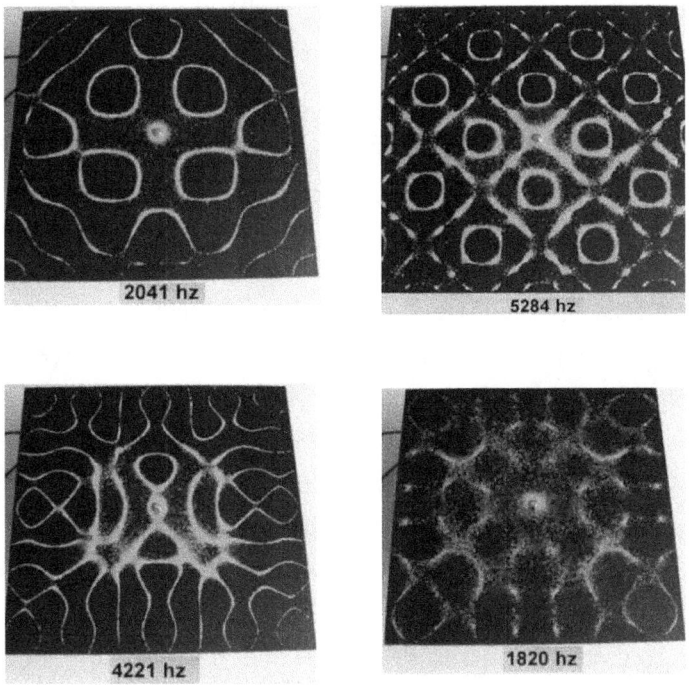

Screenshot credit: Brusspup. (2013). *Amazing Resonance Experiment.*[YouTube]. https://www.youtube.com/watch?v=wvJAgrUBF4w

Proof of making sense, having meaning, there is a *discernable design* to the different wave cycles (frequency). Here's the parallel:

Our brain discerns that the plate/sand experiment has design/meaning which is made by various notes.
~Which is analogous to~
Our brain making sense/design/meaning from all of our earthly environment.

A truly essential question to ponder is comparing nihilism to a life of purpose. Do one of these two possibilities have better evidence than the other? Is one more likely to be true than the other? Life choices will be massively affected by whether you believe there is a purpose for your existence, or whether life is random, and has no meaning whatsoever. The nihilist approach proposes that there is no point to existence. There is/are no creator(s). Everything has no meaning, existence is absurd, nothing is real. There is actually solid evidence that nihilism is a possibility. There is observable structure in our universe, but there is also, undeniable chaos and destruction. Randomness, chaos and design, all of these show up in the theoretical realm of math. Not surprisingly, randomness, chaos and design, are all in our real-world, and in our life situations as well. A Nihilist dwells on the pervasive chaos and meaninglessness of everything. Nihilism is a definite, possible conclusion when you consider the played-into-the-future, chaotic, end of our solar system. When scientists plot the positions of our planets in our solar system into the future, hundreds of thousands of years into the future, the slightest, most miniscule changes in orbit, show that eventually the planetary movement will become chaotic. Moons and planets, millions of years in the future, will have collided, or been flung out of the solar system (Batygin, 2008). This future destruction of our solar system lends credence to the nihilist view. The idea being, this

Pondering Our Existence for Practical Living

physical world, all of it, inevitably, comes to nothing. You, or the reconstituted remains of the atoms that are you, the fabric of your brain and body, eventually, like the universe, come to nothing. This scientific modeling shows our physical solar system, possesses no justification, no magnificent reason. The logical result of this line of reasoning is that this chaos illustrates there are no rules of conduct/principles/standards or social code written into the universe. Nothing has any permanence or reason, we, and our universe, inevitably come to ruin.

	Life is absurd no reason or purpose for our existence (nihilism)	There is a reason for our existence
Is there concern for humans by a god(s) creative force?	No observable interjection of a creator or creators. Creative entity does not matter in the least.	Creative entity is behind good/evil, things happen for some reason. Feels good, emotionally correct to hope for interjection.
What happens in the afterlife?	Nothing. Ourselves, and our solar system goes to chaos, sun explodes, planets skitter off. Universe is mostly chaos; our Earth tends toward rot/decay/chaos. No moral code written into the universe.	The mind of the creative entity is mimicked in this creation. Afterlife options: Reincarnation, end of game, new level, mythologies (Zeus, God Father, Allah, purgatory, heaven, hell.
Chaos or design?	Chaos is prevalent. Nature/chance/statistical, lucky that we exist at all.	Design in world = designer.
Is there an existence of a soul?	Brain is an organ, ceases to exist at death, just like our lungs, heart, foot, etc.	In the case of near-death experiences, people have brain function (seeing things/people) with no medical brain function. Some essence of ourselves carries on.
Is there a purpose for our existence?	There is no reason for our existence; reason is an individual invention. Individual body "goals" are merely evolutionary mutations. We are only an object, a thinking thing that lasts for a while on this planet, then disappears.	Individual parts of body have a goal, so why not the whole person? Our thoughts create purpose, or discover reason for our existence, a soul may carry on in some way.
Moral choices, what is right or wrong?	Justice is an illusion, what constitutes justice is in the mind of the individual as per teachings of family and culture. No perceivable justice in this life.	Longing for justice. Right and wrong, we feel emotional steering from an internal spirit.

There does seem to be more evidence on the side of there being a purpose to our existence *if* the design we observe equates to meaning, but neither view, nihilism, or purposeful life, should be labeled as impossible. Neither view is wrong, but in a discussion to find if there is a reason for our life, there is naturally, a lot more material in which to delve if there *is* a purpose for our existence. Truth, knowledge, beauty, morals, and reason itself, all become moot, or, at least folly, if nihilism is your chosen belief. Why try to even guess as to why we're here, if it all is meaningless? Plus, the preceding design evidence covered in this book, observed in the fabric of our universe: homeostasis, sound wave cymatics, mathematical/logical rules, π and Mandelbrot set. These brain-discerned, meaningful, organized structures are in direct contrast to a world that has no meaning.

Whether life has meaning or not, has a massive influence on the way you decide to live your life. However, what chosen actions would make your life seem valuable is not dependent on there being an overall purpose. It is possible for people to believe there is no purpose to this life, yet, still make choices on the basis of knowledge, social benefits, feelings, and sense pleasures. Since we are cognizant of existence in this earthly realm, why not make this the most satisfying reality we possibly can? A believer that there is a purpose, and the nihilist, both choose/complete the actions in this life that fulfill their individual reason for existence, and bring the most fulfilment. Living according to your own discernment, your own senses, this seems to be the logical approach, especially for a nihilist. To base reality on your own sensory experience, knowing you may have perception bias, means nothing, because everything is nothing to a nihilist. There is "less pressure" on choices, thoughts and actions, because everything we know, comes to nothing. Therefore, admittedly, this book will heavily lean toward the view of our lives as indeed, having a purpose. If nothing has any reason, or meaning, why read (or write) a book about why we are here, existing?

Even apathetic people make choices every day, all day. Our internally created hierarchy of what constitutes important actions tells us this existence matters, at the very least, it matters to us on an individual level.

There are people who remain only in the "just survive" level of life. Many people may never come to any deeper reason for why they actually exist. These people with no interest in delving into anything deeper than food, shelter and whatever toys/diversions (i.e. alcohol, drugs, materialism, etc.) society has told them were the 'cool' thing to be obtained. However, if there is truly no reason, if there is no soul to carry on after death, if every action we undertake means nothing, well, then, nihilistic people who stay at the unquestioning, just-survive, basic please-my-senses level, may actually have the best approach.

Chapter 10

Reason For Purpose Over Nihilism

Humans are a being that can reason, weigh options, solve problems, and decide on beneficial/detrimental actions. Without using reason, it would not be possible to question why and how we even exist. Purpose and function are used to describe how things are, and why they exist. We have knowledge of our own existence and with that, ideas of how this life *should* be. Let's use a car engine for a metaphor. We know what a car engine is, we could describe how it looks. We could say what the car engine is supposed to do. We even know *why* a car engine exists, to help us move around quickly. We all know that if we took a sledge hammer and started pounding on the engine, that's not beneficial for its function. Conversely, if we kept fresh, synthetic oil, filled the engine with quality gas, these are beneficial actions to the engine. There are "right" and "wrong" ways for a car engine to exist. Relating this to our human lives, how well do we use our ability to discern *ourselves*? What is beneficial, and what is detrimental to us as a living human being? This observation of ourselves, of people, this is the way to figure out our best way to function. Without a soul, or a purpose, a true nihilist is free to make any sense pleasing, or emotionally satisfying action about anything. Whether a person decides the evidence supports this nihilistic viewpoint, more than there being a purpose to life, is up to individual opinion and judgment. However, if we don't have any reason, purpose, or an ongoing "soul" from our physical self, well, then, we are really just an object – like a rock, or a grain of sand. A nihilist is, for all practical purposes, just a thing that exists for a while, then disappears. If this is true, our life is limited to this existence only. As opposed to if we have a purpose, or some form of an eternal soul. Having the ability to choose beliefs and actions, having a spiritual, unseen side of ourselves, inevitably makes us **not** merely an inactive object limited to this time and space, such as a drop of water, a rock, or whatever object.

Using rational thought, there is logical reason as to why there could be importance to (in opposition to nihilism) your individual life. Your human eye has lashes to protect the cornea; the cornea is an outer part of your eyeball. There's a pupil that opens and closes determining how much light gets in. Your eye has an area called the retina where light that comes through the pupil and lens, hits the retina, and those images are sent to your brain. Your brain interprets those images. The whole mechanism of the eye has a goal to attain… to see. The eye has a function, a job to do, the eye *has a purpose*.

The same goes for your ears, ears are made up of parts to perform a function, to **do** something, to hear. Our physical bodies have various other parts that are all performing an action, there's an obvious reason for their existence. These individual parts of our physical body all have *a task to accomplish*. So, now, wouldn't it be inconsistent, or illogical, to think the entirety of you, as a whole person (physical, mental and emotional) *does not* have a purpose? Even though all the parts of which you are made, **do** have a clear purpose? This is a rehashing of Aristotle's thoughts from Nichomachean Ethics (Aristotle, 2012). Since separate parts of our physical bodies, (i.e. our eyes) have a function, well, then, all the parts together, the whole person, must have a function. Can we logically be made of separate functioning parts that have individual purposes to accomplish, yet the entirety of these physical aspects of our selves, do not have a purpose? If we settle on having a purpose for our existence, this leads us to the possibility that there might be a reason outside of our physical selves.

The brain is an organ, just as our heart, kidney, liver or whatever. These physical organs die with the physical body. They cease to exist. To this point, along with our chaotic universe, the nihilist has a strong argument. However, opposing our physical annihilation, the brain and nervous system of a human, includes a specific nerve, called the vagus nerve. It is located in the area of the upper chest, near the heart. This is the physical location where we *actually feel* our emotions in our body (Perelló et al., 2022; Johnson & Steenbergen, 2022). Is this sensation an indicator of the location of our soul? Is it possible, the emotional feeling part of ourselves carries on beyond our physical body's death? Or, is this vagus nerve just a body part that ceases to exist when we die? This is where the most important aspect of our existence comes front and center, our emotional side, our affective domain. Our feelings, the reaction to the stimuli we experience, this is a real experience for us individually. These feelings direct our actions and could be another indicator of our purpose. Everyone knows the experience of the tug of our feelings, the pull of what feels right/wrong occurs in each of us individually. This true, emotional experience, is actually knowable evidence, from directly within each one of us, that we have a reason for existing. We feel it. These inner urgings are the indicators that tell us what to do, and to say continually.

Researchers at the University of Virginia have multitudes of evidence for spirit/soul existing with or without the physical body (Greyson, 2021). When people have no brain activity in a near death experiences, how can they see visions? Many recount, with absolute accuracy, what happened in the operating room while they were watching themselves on an operating table (Greyson, 2021; Tucker et al., 2017). How can this be when there is no brain activity? Only through a separate part of us, distinct from our physical nerve system, this is the only way out-of-body, or near-death experiences are able to happen. It is not a far-flung hope that when our physical self dies, some spirit/soul, separate from our physical selves, carries on. Near death experiences are clear evidence of some aspect of ourselves, that is separate from our physical body (Tucker et al., 2017). Earlier in this book I went over other proof that there are brain waves, outside, physically separate from, our biological brains: 1. Scopaesthesia, 2. brain-computer interface (to physically drive cars and move human limbs) and 3. inventions simultaneously occurring around the globe with no communication. These three points, along with near death experiences, are knowable proof that we can observe, that 'some essence of us' could exist outside our physical brain and body.

If we do acknowledge that we probably do have a purpose, if a human being has a purpose, then, what is it? Lots of ideas have been put forth in the past. Is our purpose to worship some god or gods that made us? To maximize our own pleasure? Is our purpose to feel good? Merely to survive? To feel emotional happiness and contentment? Are we made to entertain some creator, or deities? Are we here to help other creatures? There are many possibilities.

Chapter 11

Creators, Other Possibilities & Functionality

The ideas presented thus far lend credence to the possibility of a designing entity:
- Mathematical/logical design (even our concept of design)
- Iterations of consciousness
- Free will
- Mind cloud/universal consciousness
- Inner, emotional steering
- Longing for what we see as justice or "right"
- Time/space
- Brain waves (spirit/soul) outside the body
- Individual physical parts of the body have goals, so our whole body must as well
- Design/Order intertwined with chaos
- Equilibrium / homeostasis

For the purpose of ease of discussion, whether there is, or is not, a creating force, (God, or Gods) we need to have a name for that possible entity. Hereto forth, will be called "Creative Entity" in this book.

Pondering Our Existence for Practical Living

If there is a creative entity, a consciousness, or God-like beings, it is doubtful that this entity is a solitary "Almighty Father." A long-bearded god, whether it be Zeus, God, Thor, Heavenly Father, or Allah, or whatever name that any story-telling mythology has passed down to us, probably does not exist. We see this from our own human creativity. More likely, the Creative Entity, if it does exist, is unfathomable to our mental abilities. It is possible there are infinite creative entities, that created the next iteration of design. Math formulas have successfully described many physics phenomena of our universe. Is math/physics male or female? Does math have a caring heart toward anything? The physical rules are set in place by this possible Creative Entity.

Let's consider our human endeavor to create entertaining video games. What happens when a character in the game dies? That character in the video game comes back to life and starts over, or, they graduate to some "next level," and continue. Our human race, human minds, created these game scenarios. Is our human creation of video games a clue as to what happens in our own death? It is a possibility that **we** come back to life. Where did the makers of games come up with this ubiquitous idea? The comparison could also be carried to earning a "next level." What if the ancient Hindu belief about karma; actions in this life have a bearing on your future existences, what if that really happens? I am not saying this is fact, (I will discuss karma and eastern religions later) rather, that it is possible. And at least we have a template from our own creative thought, which is, in turn, a template of a possible creative entity's thought.

If there is a creative entity that made us, it is more than likely plural in nature. Is there one human creating advanced, artificial-intelligent robots? Is there one human creating all the games with death, and rebirth, and graduating to the next level? Obviously, many humans contribute to creating video games, and also the creation of artificially intelligent robots. Therefore, it is logical to think there are 'many' that comprise the "Creative Entity" that possibly made us. Currently, there are advanced robots being created, robots which might be given the ability (by us humans) to reproduce themselves one day. No, there is not one human working on these projects. There are many, hundreds, more likely many thousands of scientists, from varying fields, working on these intelligent robots. If these scientists succeed in creating a robot capable of decision making and reason, and then, we give these robots the ability to reproduce themselves, voila, you are witnessing how we, ourselves, came into being! Us humans, a "created something," with the ability to reproduce ourselves, is analogous to our intelligent robots, being created by humans. In practical terms, we would be "Gods" of that race of robots we created. Creation, within creation, within creation. If the Creative-Entity-robot-creation comparison is valid, these intelligent, copied robot-beings would emulate us, and have our implanted characteristics. The robot mind would be what we know from our own mind. The human race would be as we vision the "god race" say, of the Olympian Gods or a Heavenly Father. The parallels are obvious to our own existence. We are the "A.I. robots" to the Entity that created us. Would we, the creators of this future robot race, be involved in all of the robots' individual dealings? Would we care, or intervene, if one created robot was violent toward another created robot? Probably not, and this is exactly what we perceive in this life.

As of now, there are well-attended and televised robot competitions. Some of these competitions are based purely on one created robot fighting and destroying another robot. Humans, being the "creating gods" of the robots, prepare and plan these robot battles. Isn't this a fitting parallel to Homer's Iliad? Quite possibly, the capricious, ancient Olympian Gods who entertain themselves by interfering in the affairs of people, is a metaphor for the functioning of our reality. The appearance of things "going our way," with a "god on our side," or, conversely, "hitting all kinds of road blocks," does seem to be a perfect summary of our experienced reality. Even in these robot fighting competitions, the creators make their robots, and send them on their way to their "life" in the combat area.

Chapter 12

Do God(s) Intervene?

A key point about the human existence on earth: No Creative Entity, God, or Gods, are intervening in our lives. If we are the creation of some intellectual entity, that creative entity cares about us in the same way we think about the robots we create. Whatever Entity that possibly made all of this perceived universe, *does not* intervene. The physics rules are set in place, this entity is obviously indifferent to our individual perception of joy, or suffering. There are countless unjust, horrible, situations now, and throughout history, where no creator showed themselves to destructive abusers.

Let's take the case of some blatantly, obvious situation that most people would see as egregious wickedness… a school shooter. In 2012, a man shot and killed twenty-six people, including twenty children between six and seven years old. In Newton, Connecticut, Adam Lanza went into an elementary school, and shot twenty children and six adults. In one classroom, Lanza murdered fifteen, innocent, defenseless, first grade students (Onion et al., 2012). Lanza murdered the children for no rational reason. If there was a "just" and "good" God in control, helping the innocent, as many religions teach, that would have been a pretty great time for a divine intervention. No intervention came from Allah,

Zeus, Heavenly Father, Brahma or Vishnu. Quite the contrary, Lanza "was allowed" to kill five more six and seven-year-old children in another part of the school. The reality is, the shooter wasn't "allowed," there is just no creative entity concerned about us, at all, individually.

There are many such chaotic, dreadful, incidents, similar to an elementary school shooting, that occur every day, all the time, all over the world. From very young, to very old, unfair, "bad," incidents happen constantly: cancer, bear attacks, car wrecks, tsunamis, tape worms, rape, murder, etc. Would not these types of incidents be a wonderful occasion for a loving, just, caring, all-knowing, God to step in and take action? If a caring, loving, watchful God cared for us, why would this creator make other creatures that harm us, or even eat us? Or insects that sting and bite? A caring, loving, watchful God would not create pain, suffering and discomfort for his loyal followers. What should a "good," omniscient God to do? If your belief is that there is an all-loving God watching you, caring about you, you are certainly deceiving yourself. In the cases of all the constant injustice we see around us, there is no possible, logical way to say a "good creator(s)" is/are watching over and caring for us.

Since we see heinous evil and injustice, it is illogical to believe God has the ability to and/or God wants to, eliminate our view of something "evil." Here is ancient, Epicurean logic: If there is an all-loving God that hates evil, this God-like entity would stop it. Since we do experience incidents we individually perceive as "evil," it is possible there is some God-like entity that *can* eliminate evil, but does not want to. It is also a logical possibility that this god entity wants to get rid of evil, but doesn't have the power to do so. It is likewise, a logical possibility that there could be a god-like entity that is powerless to get rid of evil, *and* does not want to get rid of evil. If we use our five senses, and look at the reality of our life, stripping away the traditional, and/or the pervasive teachings of our family and surrounding culture, we come to a more honest perception of our existence. If an all-loving God exists, this entity would ***not*** be loving at all, in fact, this god would be sadistic and deranged to allow innocent people to be diseased, hurt, murdered, etc.

Many people do manufacture faulty excuses for an all-knowing and loving God to allow these horrible acts. Some people say God does listen, and then give many rationalized excuses for a disinterested God: "God is with you through the tough times." "God allows you to go through trials to refine you." "He's watching us from a distance." "The Creator is carrying us, or walking with us." "God wanted to teach me a lesson." "I wasn't ready for the answer yet." "I doubted, so God didn't give it to me." What a sad state, people blaming themselves, owning shame and guilt, all for the failings of a fairytale God. This is the Santa Claus-type God that causes undue, unwarranted shame and guilt in individual people. If believing in some God, a God who is really wanting you to have a happy and fulfilled life, why wouldn't that god give his loyal subjects a few wins? Heck, why not give lovely, good and pleasurable experiences all the time? Looking at it from the opposite direction, why would an all-knowing, all loving God, give people 'bad presents' like accidents, sickness, and/or the allowance to be physically hurt or ripped off by others? If an omniscient God allows any and every 'evil' to occur, then what good is it to pray? Or to even follow such a being? If believing in, and praying to God doesn't change your practical reality, then why do it? Conversely, changing your own thought process and behavior, that can change your practical reality, this may actually be beneficial. The observable fact is, whatever creative entity that may have made this creation, that Entity is not listening to anyone's prayers. A prevalent answer/excuse would be to push justice to some afterlife judgement scenario. Thinking, "Our reward comes in the afterlife." So, there could be hope that a creator(s) all of the sudden will decide to become active in an afterlife scenario. Some people think our creator(s) keep track of these actions, and justice will be doled out in the next existence, after death. This detrimental idea can hamper effective actions that could happen now, in this life.

There is another excuse often given, that bad/evil events happen because humans have free will. Many humans rationalize, "It has to be this way because humans need freedom of choice between evil and good actions." This also makes no logical sense. Why must humans' having free will be united with a loving God never intervening? There is the possibility that humans have free will, yet, can also be corrected (or have any god intervention) by an omniscient God **during** this life on Earth. In fact, most major religions believe this correction of wrong doing, karma in this life, or a reckoning that happens when humans die. If God can have a reckoning in the afterlife, or correct evil and injustice in the afterlife, then corrections along the way in this life should be/would be truly effective indicators that a caring God is real. The reality is, there is no correction. It is absurd for humans to decide right and wrong, but also, concurrently, have zero Godly intervention. Free will and visible, Godly intervention do not have to be mutually exclusive.

I was about eleven years old when I made a "Jesus Corner" in my closet. I was brought up Catholic and I don't remember if a priest in a sermon, or a nun in a catechism class, told us kids to make a corner in our closet devoted to God. Yes, I knew it was sort of odd, so I didn't announce this to everyone. I had mass cards with prayers in my closet "Jesus Corner." It was a middle shelf, so as my shirts hung in the closet, they camouflaged my corner of devotion. I actually did go in there and pray sometimes. Emotionally, I felt good in there, I had a mystical, special feeling. Herein lies a noticeable benefit if one does pray, it could be a form of meditation that brings peace to some people… Even if it is ineffectual in practical living. Now that I'm old, I don't remember my exact words or prayers, but I know what my thoughts consisted of, because my older brother Richie, found my Jesus Corner one day. On the back of one of the church cards with a picture of a saint on the front,

I wrote my own little prayer. I know I asked for all kinds of gifts in this written prayer, I specifically remember asking for a million dollars. As older brothers will do, Richie had a field day making fun of my prayer. "You don't ask God for money! God isn't there just to give you what you want in life." Well, why not?

The reality is, we all lead our lives as if no one is watching, because that is the truth. Think of all the things you do in secret; the "unholy" embarrassing thoughts we have every day of our lives. Does this creator, or these creators, care for us individually? The evidence is not there. No creator(s) are watching or caring in the least. Try this, curse your God, see what happens. Try this, praise, and love, your God, and see what happens. There will be no change in your existence in either case. This is an observable reality. There is so much suffering and injustice; a fair creator or creators would step in to stop rampant trampling of people, both the literal, and metaphorical trampling.

Apart from any irrational excuses made for an inactive, all-loving, all-knowing God, it is plainly evident that no god is intervening to stop heinous violence and injustice. Still, there is strong evidence that there is some sort of Creative Entity. However, the Creator(s) do not miraculously intervene, even when "great evil" is being committed. The way god, or gods work, is to set our realm in motion, and it is just rolling along. The world keeps spinning, oblivious to our individual experience of depression and/or triumph. Horrible incidents do not occur so that we carry on with some botched plan. This realm is a machine, an interlocking, designed and functioning mechanism, with no intervention from a supreme entity to determine outcomes this way or that. Our realm can be logically understood by mathematical equations; these physics rules are/were set in motion by some creative entity, and we are just naturally rolling along as part and parcel of this creation. The absence of godly intervention speaks loud and clear. Wouldn't it be wonderful to hear a clear voice from heaven? God's silence is deafening.

Pondering Our Existence for Practical Living

Maybe we should be grateful we don't hear some Creator's voice; the absence of the audible voice means something big. Earth keeps spinning regardless of our ranting, raving, pleading and praying.

Chapter 13

What Constitutes a Well-Lived Life? If There is a Purpose, What Is It?

Let's go back to Aristotle's physical eye example, to hopefully parse out a real reason, a true purpose, or purposes, humans may have. Since the function of the eye is the same for all eyes, is the function of a human the same for all humans? Since every person's ears have the same goal, do all humans have the same goal? The eye does not seem to have a choice in its function – it just does what it is supposed to do, seemingly, in an automatic way. So, what is the "automatic" in the functioning of a human being? Therein lies the natural function we *are supposed* to carry out. I am specifically thinking about babies, yes, from the minute a baby comes forth from the womb, even in the womb, babies are subjected to other humans' influence. However, some things babies do seem to be automatic, without any training or coaxing from parents. Observing non-indoctrinated, new-born babies could be the key for how we should live our lives. What does a new, "untrained" baby, unconcerned-with-other-peoples'-pressures, or cultural indoctrination do?

Pondering Our Existence for Practical Living

Babies automatically move their face and lips to "root," to look for food, to satiate their hunger, this is one automatic action we are meant to do. Babies communicate discomfort, or their need of something, by crying. So, another, natural, important goal, is that we are meant to communicate and seek comfort. Babies suck their own fingers for pleasure. Babies pass gas and then they smile, or at least have a pleasurable facial expression. Basically, a baby's actions are bent toward trying, longing for, looking for, and achieving, a way to please his/her senses. Babies automatically try to avoid pain; they go toward pleasing their five senses automatically. This is key, we humans only know we even exist because of stimuli through our five senses. So, if we look at babies, seeking pleasure is a natural/automatic goal throughout this existence, it is a correct way to live. Corresponding to pleasure seeking, babies avoid pain as well. Nine-month-old babies have been shown to fear and give attention to hissing snake sounds, angry human voices, and crackling fire noises. All those reactions are built in, without any teaching by an older guardian or parent (Bower, 2013). All of my own babies, right after being born, were pricked on the bottom of their foot, and blood was squeezed out to check for metabolic conditions. All my babies cried and pulled their foot away from the painful prick. These automatic actions, to avoid pain, to seek pleasure, are automatic goals for humans, the ways we should live, and give reason for our existence.

Researcher, Harry Harlow, designed a surrogate mother experiment. Infant monkeys were taken from their mothers and given to two "substitute mothers." One mother was made out of wire and wood, and the second mother was covered in foam rubber, and soft terry cloth. The infant monkeys spent significantly more time with the soft, terry cloth mother than they did with the rigid, wire mother. When only the wire mother had food, the babies would feed and then immediately returned to the soft mother. Soft comfort through our sense of touch, is a goal for us, a natural, and automatic, correct way, to exist (Harlow,

2018). Human babies need touch for development, it is true that the death of an infant occurs when babies are not physically touched (Montagu, 1986).

Babies also search out comfort, companionship, in an automatic way. Research on infant rhesus monkeys was completed. Some rhesus monkeys were taken away from their moms and raised in cages, away from monkey peers and their mothers. In social isolation the monkeys would stare blankly, circle their cages, and self-mutilate. Many refused to eat and died (Suomi, 1971). Physical touch, cuddling, love, and communication, these are all basic, automatic goals for young monkeys, as well as us, as humans. This is what we are supposed to do, what we are **meant** to do. It is so simple, yet so profound, these automatic goals we see in newborns, are the natural way for us to function. The goal of feeling love, and affection are a part of our purpose for existing.

The acquisition of knowledge is another basic, automatic, reason that we humans exist. Gaining knowledge, the exploring and learning of the baby's environment, learning about our world, this is automatic (Tamis-LeMonda & Masek 2023; Gopnik, 2010). Newborns, over the first weeks of life outside mother's womb, start to examine their own fingers and hands, they study their own movement. Young babies will grab anything and put it to their mouths to see how it tastes. Observe how a baby studies and reacts to a family member's face, see how a baby tracks a rotating mobile. Learning is natural and automatic; learning is as natural as rooting for a breast. We automatically try to "figure stuff out," we humans are naturally meant to learn.

What "un-indoctrinated" infants do, naturally, and automatically, this is what all humans are meant to do. Our way to exist is to please our senses. These natural actions are the simple, basic, reason for our existence. Happiness is paramount, ask anyone the indicators of a great life, nearly everyone would say, "To be happy." So, why don't we strive to minimize pain and maximize joy? Joy and happiness coming from pleasurable sensory experiences, this is the way to a fulfilled life. Is it not natural to value experiences that bring fulfillment? So, it follows, that we should do these automatically pleasing behaviors. You'll be happier because you are doing what is naturally meant to occur. We are functioning as we are meant to function. I currently have a dog named Lucy, she is half hound dog, and half Labrador Retriever. We walk Lucy back to a patch of woods behind our house. No one ever taught Lucy how to hunt mice. It is incredible to watch; she leaps high in the air and pounces on the area where a mouse hole exists. This pouncing-tactic allows success for Lucy to catch a mouse. No one ever taught her this technique; she's lived with only our human family. When she performs this untaught, natural leap-pounce-catch-a-mouse method, she is so happy. Lucy is doing what she *is meant* to do. This brings her pleasure and happiness. In the same way, we need to do the things that we are naturally born to do.

In addition, to seeking these automatic actions to seek pleasure, these preferences should not give us a sense of guilt. Why are so many pleasurable actions a damnable offense for so many religions around the world? So often, religious tradition, family teachings and cultural pressures, make it clear we should feel guilty, for feeling good. Guilt, self-induced pressure, fault, worry, are all unhealthy. Why do so many people choose to take on guilt and shame? One reason is that people enjoy the ready-made right-and-wrong put forth by others. Take for just one example, women of the Hindu religion, when they are naturally, inevitably, menstruating, they are not allowed in temples, they are not allowed to cook, or eat with others. Women are not allowed to wash their own hair, or to swim, and are secluded in huts. This cultural view of impurity is restrictive and stigmatizing, even if it is culturally accepted (Maharaj & Winkler, 2022). Guilt and shame for something as natural as menstruation?

Another reason people take on surrounding pressures to act and think a specific way, is that it's easy to accept consistency and regularity. Accepting without questioning has security built in, "This is just what we do. This is just the way it is." In addition to this, many people enjoy controlling others as well. Some people love power, and truly enjoy controlling others. It's detrimental that people feel guilty about natural actions that could truly bring them joy, but it's especially detrimental when people feel guilt over their own, natural, biological functioning. Importantly, the way we look at these experiences, and the expectations we set for these sense experiences, has a direct impact on our self-rating of how happy we are. There are people abiding in incredibly beautiful, seemingly perfect (from the outside view) circumstances, that are still extremely unhappy. Internal choice of emotion is difficult to steer when awful circumstances come our way, but our reality is always designed by our own mind's interpretation of any, and all, events of our existence. Even if there is pressure not to, when we please our own senses, hearing, taste, sight, touch, and sound, we find happiness and tranquility easier to attain. If you emphasize focusing on joy with simple sensory pleasures, tranquility and happiness will emerge.

Millions of people, from all parts of this globe, every country, come to Yellowstone National Park in the U.S.A. They spend thousands of dollars, renting cars, buying plane tickets, buying lodging, time planning, driving, all in the hopes of seeing some certain animal. Maybe a bear, a wolf, or a ram, or a moose, or whatever. Let's say they get lucky and see a moose that they wanted to see. That moose *is* amazing; it *is* magnificent; I'm not detracting from the incredible moose. However, think about it, that moose isn't *doing* anything. The moose never solved a simple math problem, ever. It didn't create anything. It's living as it is meant to, as human babies do. The moose eats, sleeps, poops, and makes baby moose, yet… it is incredible! Why? Here's why… because it exists. The cells of that animal are arranged in the beautiful order of a moose. Well, humans… we also exist! We already made it! Just being alive, we already "made it." We humans, in this respect, are no different than the moose. Our natural, automatic actions, are the main, basic reason we exist. These natural actions we are meant to complete, give us contentment. Plants turn sunlight into food, animals/humans seek food, shelter and procreation, these actions are natural, automatic, they are "right." When we take care of our basic needs, those actions bring tranquility and decrease our sense of suffering, this could be all that is needed. We are all just as glorious as any other creature.

Chapter 14

Picking a Secondary Purpose, Beyond Automatic/Natural Purpose

Does anything about a human make us different from plants and animals? Is it our ability to use logic/reason? Is there a difference from other animals that we have these feelings of right and wrong? Ants, bees, and many other species have a perceptible order to their own society. We can see each, individual ant bustling around, and completing tasks for their society. Is pain and suffering bad? Is feeling joy and comfort good? Is there something beyond gaining knowledge and sensory pleasure? The truth is, we humans certainly do complete tasks beyond the automatic. If the goal in life is happiness, the pressure to be more, to accomplish more than our automatic goals of pleasing our senses, this could add stress and anxiety. However, this striving could also lead to greater happiness if/when we work toward achieving our goals. In determining a "best way" to live life, sometimes we put pressure on ourselves. At times, everyone becomes sad, lonely, or lost. Even a person living, physically, right next to other people, may feel lonely, lost and have difficulty finding goals, or reasons for their life. Siblings feel pressure to live up to (whether it be real or imagined) accomplishments of their other siblings. Sometimes, there is undue and unrealistic pressure from parents to push their child to attain certain goals. Many people are pressured into "a best life" as determined by their parents, or others. The chosen goal, if not truly longed for, may become the biggest source of stress. Desire, competition, comparing yourself to others, definitely leads to stress. The temporary victor may feel happiness, but it will be ephemeral. If/when a person does "win," there will be the stressful urge to do it again, or achieve greater heights. Most people do pick other goals, or reasons to give purpose, and meaning to life. When we assign importance to these chosen, life-fulfilling actions, this longing will increase stress. Accomplishing goals may add stress, but it may also add great pleasure when the goal is attained.

Let's look at the ignoramus, a person who fails to reason. A person who fails to regularly use their intellect, to gain wisdom, to make wiser choices. These people usually have what is often viewed as hard/difficult, and/or "unproductive" lives. As stated earlier, the automatic, un-indoctrinated actions babies perform, are the "first-tier" goals of learning and sensory pleasure. Humans have already successfully "made it" if they live in the first-tier. However, to remain in first-tier only, these people are usually viewed by other people as going through the motions, or even making, sometimes, non-beneficial choices. Sometimes, the basic person winds up giving their lives to some vice. The ignorant, apathetic people, are often at the whims of the driven, goal-in-life oriented people. Even smaller, everyday errors in using one's ability to make beneficial choices, will eventually result in less freedom and less power over one's own life. Therefore, the virtue of using reason, rational thought, your intelligence, pursuing goals, is good. This aids in a more fulfilling life. Afterall, gaining knowledge is one of the "automatics" babies do. Conversely, failure to reason, or failure to strive for being better, (at anything) can be viewed as something very negative, and detrimental to a person's existence. There is superiority to a life with intellect, and adding secondary purpose, to have a reason beyond basic natural sensory happiness. It's a possibility of ours, some would say a responsibility, that in order to have a fulfilling life is to choose second-tier goals for your existence. So, what is your meaning? Since there seems to be a more fulfilled life when people live with an additional reason/purpose for our existence.

Clarifying here, are two levels of goals: 1. Automatic, natural purpose/goals for your best existence. This includes the natural, automatic actions we do, procuring food, shelter, gaining comfort, looking for ways to bring pleasure to your senses, learning and communication. These are the things we automatically have as goals set for us when we are born. It is *possible* that a good life can be experienced with just these first level goals. 2. The second level of goals, "second-tier goals," for human life are a chosen purpose, these are additionally chosen goals for a fulfilled life. The overarching ideas to pick from, for these secondary level goals, are purely an individual decision. How does one know what secondary goals to pick? Do you long to rack up accomplishments? Earn awards? Do you want to perform good deeds? Or any deeds of notoriety? Do you long for love? Relationship excellence? Do you yearn for just having experiences, memorable "peaks" of emotionally charged incidents?

Second tier goals for life will vary from person to person. Any second-tier goal is possible, and none are wrong. Naturally, these secondary goals are found in completing what activities you love to do. What activity do you enjoy so much you completely lose track of time? In the secondary level of purpose, we choose according to our own values and meaning for life. These second-tier goals are our choices about what is important to us. The thing to remember is, there are no wrong answers. Focus on a musical instrument, athletics, piling up material possessions, science, serving, art, leading etc. You create the reality of what your purpose is to be. These second level goals may cause stress, but these chosen goals also come with reward of accomplishment. Listen to your internal steering and do not let the world make you feel guilty about pursuing whatever actions you truly enjoy. These second-tier goals act on our lives as the wind acts in our world. You can't see the wind, as you can't see the internally chosen second-tier goals for your life, but you can feel it, and see its effects on a life with purpose. Similar to the wind, your choices in second-tier goals may (and probably will) change over time.

There seems to be evidence, and some advantage to a more fulfilled life when it is based on reason and purpose, rather than our existence being an absurdity (Sosin & Neubauer, 2024). A person with secondary goals beyond survival and sense pleasure, people who have a "second tier" chosen purpose, will have more life experiences, and a diversity of life experiences. The person searching for reason, attempting to parse out *why* the vagus nerve gives us the impression of emotion which guides us, this learning-person is *less* likely to be stuck in the metaphorical rut of a life of mundane work, and a maximum number of drunken weekends. There were adolescent boys who were visiting my office, talking to me one morning. The conversation went to the question of what constitutes a person having a great life. One young man proudly stated, "I'm going to get a job working construction, then I'll have money to lay around and get drunk every weekend!" A declaration of a goal consisting of drunken weekends, gives a knee-jerk thought reaction of "No, there must be something more."

If the nihilist is correct, everything is fatal and frivolous, this life could be all for naught. In this case, the nihilist is free, since life means nothing anyway. Also, whatever a person picks as their second level/chosen purpose, will probably be viewed as frivolity to another person's viewpoint. Let's just say your second-tier goal, is to play the piano well. Pretend this is your chosen purpose for this life, to play the piano. Some people will look at that goal as, silly, or useless. Others will say that playing the piano is a great and lofty purpose for a life goal. There is a multitude of choices as to what secondary goals could be chosen: cooking, welding, teaching, doctoring, throwing a ball through a hoop, making backpacks, going backpacking, making computer software, growing crops, etc. All of these are foolishness to some, and at the same time, highly important to others. Who can say one is superior to another? Fascinatingly, choosing a reason beyond automatic goals is beneficial, and simultaneously, it can be viewed as frivolity. Write it down, say it out loud, why are you here? What are you shooting/aiming for? Even if this existence comes to naught, as the nihilist believes, you at least have something that is important to go toward while we are existing here in this life.

It is possible that the whole community may be elevated when the community is comprised of purposeful living people. When people choose a second-tier goal, beyond that of what babies do automatically, all culture will benefit. When people say a life with purpose is "blessed," "lucky," and/or "right thinking," when other people admire and wish to have the life a person is leading, these are indicators of living a fulfilled life with a second-tier goal. A life worth honoring seems to be viewed as a "good life."

You pick your purpose. You pick the qualities you want to emulate, you pick your second-tier goals, and you decide what actions to complete every day to fulfill your chosen purpose. Do not feel guilty about actions that support the second-tier goals for your life. Guilt is an unhealthy weight for something you did in the past, and over which you no longer have any control. Guilt is wrapped in others' opinion of what **you should** choose as being important. Even in the moment, daily decisions, no matter how horrible, or enjoyable, there are reasons for the choices a person makes. Focus on those reasons for your choices and actions, and throw away any guilt. Guilt destroys your chance at any happiness; guilt and shame stymies bold choices moving forward.

Pondering Our Existence for Practical Living

Whether these choices involve big life direction trajectory, or everyday actions, do what you truly believe is beneficial to you. Your choices are your own, they should not be mandated by others' judgment, or others' chosen priorities. Never let guilt about your life choices weigh you down, the arrow of time only goes forward in this life.

It is fine to have becoming wealthy, or famous, as a person's second-tier goal. However, being rich, and/or famous, does not always lead to happiness. Lots of wealthy, famous people are so unhappy, some commit suicide: Robin Williams, Edwin Armstrong, Anthony Bourdain, Chris Cornell, Adolf Merckle, Peter Smedley, Paul Castle, the list goes on. Monetary wealth is a created fantasy most people think is a worthy goal around the globe today. People don't need wealth, the reality is people want what wealth can buy, which is freedom, power and comfort. What a person *does* with the wealth is what makes a person happy. The same goes for being famous, you could be famous for negative things and be quite unhappy. We need to be analytical about what to pursue in this life. If money, or fame, is a chosen goal, go for it, but realize it is truly what fame and money bring is what people long for: ease of life, safety, influence, special treatment, and/or attention.

Many years ago, I was friends with a guy by the name of Robbie. He was a great trumpet player. He was an incredibly gifted musician. I was working with him when he was reminiscing with me about the days when he played in the drum and bugle corp. People who are truly passionate about marching band can travel with corps groups up until twenty-one years of age. It takes thousands of dollars to join, travel, and perform all across the USA, and even sometimes internationally. Anyway, Robbie told me, "Steve, I was lying in an equipment van, rocking along with the bumps in the road, and I felt like I was right, exactly where I was supposed to be. I felt such contentment and fulfilment." This was Robbie's description of his emotional, internal, harmony, because playing his instrument in the drum and bugle corps was exactly what he was supposed to be doing. Here is just another example of when our feelings, our emotional reaction, steers us. These internal feelings clearly tell us, "Yes, this is right where I am meant to be." Or, alternatively, "No, this is not what I am supposed to be doing." It could be something simple, like living simply in a beach hut brings fulfillment. It could be changing the world through analyzed thought could bring fulfillment. Whatever it is, humans individually *feel* what is right emotionally.

Chapter 15

How We Decide What is Truth

When I was very young, I loved to draw with crayons, creating art scenes. I was up late, far past the time my parents wanted me in bed, so, I only had a very dim, light from my closet, as I was sneaking drawing time. When I was drawing my pictures in the dim light, the colors I picked seemed perfect. The crayons were all broken and unwrapped, there were no paper labels on the crayons in the stray crayon box. These night-drawing sessions happened many times, because I thought what happened was so mysterious. The next day, in full light, the colors of the pictures were not the way they looked in dim light. The colors of the pictures were bungled; the picture was really weird looking. The colorful art had a whole new, and very different, look in full light of day. My perception in dim light was different from the truth of my perception in bright light. This change in the truth of our perception does not only happen with colors, these changes happen with every belief in observed "facts." If I never saw my drawings that were completed in dim light, in the bright daylight, I would always think with my perception, that the colors were correctly drawn. Whether I am viewing the colors in dim, or bright light, the colors are true in each, separate setting. The changing colors are truth to you, even though the colors change in different settings. Your current perception *is* truth to you. Even though it is very likely not objective truth. The "obvious," "right path" to you, is only obvious to you. The teachings we get, the surrounding culture, and your choice as to which teaching is most up-to-date, truthful, science-backed idea, you choose to believe, is only ***your*** truth. Your obvious, or "right" choices are only obvious because they are your colors, the colors that you see.

Have you met people (I personally have met many) that say they know it's true because of a feeling they get in their heart? Of course, this type of statement is scientifically unverifiable, it can't be logically reasoned. However, a person *knows* something because of a feeling they have inside. Sometimes this feeling is described in the form of a warm, confirming God of the universe, or described as an intuition. This method, does seem less trustworthy than academic reasoning, or scientific study, but if we are honest with ourselves, this emotional reaction to information is powerful, and always present. In many cases, emotion is the overriding factor in a person's decision to believe something is true. This is the predominant determiner of which beliefs are followed. Even to a "cold," factual, true scientist, proving some observable scientific reality would be lying if they denied the feeling they get from believing something they emotionally "know" to be true. The feeling of truth can come from lots of scientific study and observations, so you just feel, in a statistical sense, that something is true. Even though many scientific "truths" have been debunked over the course of history.

Reality matters less than people's internal peace of mind, this is where faith comes to the fore. Truth is in the eye of the beholder. People believe what they *want* to believe. Our emotional pull is far greater than our academic, or logical tug (Goel & Vartanian, 2011; Jung et al., 2014). Even scientific study is run by emotion and a faith that something is truth. What we want to see, expect to see, what we choose to even test/observe scientifically, is *emotionally* directed by our beliefs. Your internal thoughts and mindset are powerful, affecting your life path and your health. In most religions the god(s) has/have an influence on the actions people undertake. The truth is, this providential, internal, spirit tells people exactly what they themselves want to hear, and to do. Right and wrong is felt in the heart, it is highly influenced, but not ruled by, your environment, the latest science studies, and religious teachings and/or civil laws. The heart, our emotional reaction, is inseparable from the decision-making mind.

Where do we get our idea of what truth is? What is "good," or the idea of what a "correct" and/or beneficial action to take? Our decisions are obviously influenced by our surrounding culture and upbringing. Familial and societal traditions, and cultural pressure, act as unifiers for family and society. This steers our actions to fit in with what is considered "normal," "correct," or "true." Derision is not intended for this powerful way that many people understand what they deem as truth, but there are limitations and false beliefs from just owning what we have been taught. Since we are partly designing (in our minds) our own reality, accepting predominant local teaching is legitimate. These beliefs are passed on by parents to children. These dominating, overwhelming, accepted local beliefs, also from some big organized religions, or smaller cults, guide many peoples' lives. If you are living in one part of the world, what is "right," "good," or "truth," will vary greatly from some other area, but still, local culture is highly influential on everything we do.

Many will say their scripture, their god-breathed scripture, has the true, and the right way to live. Yes, well, that depends on the beliefs and the geographic area in which you reside. What is right and wrong changes from person to person, culture to culture, and even individual situation to individual situation. If you were taught/indoctrinated, since the time you were born, that women should wear a head covering in public, then, women would wear a head covering. If you were taught people from a given ethnicity were sub-human, an un-evolved animal that deserved no rights, you would accept that as truth. If you were taught that Earth was the center of the universe, you would accept that as truth. If you were taught that it is right to put others' needs before your own, to be kind, you would accept that as truth. If you were taught that women were to follow the orders of men, your husband was doing his duty to burn you with acid if you didn't obey his every order, you would accept that as truth. Whatever your take is on these wildly varied teachings, these are all *real teachings* from someone's "truth" or "scientific proof." I would predict that for each of just these five teachings (women have to wear head coverings, racism, the geocentric model, living humbly, wives' subservience to their husband) you had an emotional reaction. Another kind of emotional reaction is felt when I describe accepted and/or familiar teachings, such as caring for others. Emotional reaction steers us as much as any rational thought, or scientific understanding. What is "sin," and what is "righteous," changes according to where you live and what has been taught.

Another way that people come to decide what is true, is by logic and reasoning. Looking at facts and deciding on connections. The first example is from earlier in this book, that the parts of the body have goals to complete, so the person on the whole, has a goal to complete as well. Using logic is a legitimate approach to truth, our mind linking observable occurrences have made major breakthroughs in the betterment of people's lives. Another example of the beneficial use of reason: someone, in history, used reason to create a soap that is able to isolate germs/particles and wash them from your hands. As a result, humans now live longer lives. A third example of people using reasonable conclusions from observation to better human lives, is the ever-present Global Positioning System (GPS). Einstein's Theory of Relativity was the first step toward all of us using electronic navigation (Buenker, 2014). Reason from observations, extrapolating information, to figure out truth, is highly effective. Reason and observation lead directly to another, very popular, way to come to truth… science.

Many people will, especially currently in the United States, declare that science is truth. Science is another valid way to come to discern reality, but there are caveats. To observe what happens, using mathematical calculations, has taught us an incredible amount of what our existence means. The problem is, science is not immune from its surrounding culture, and science "truths" have been debunked throughout history. Up until the time of Einstein, scientists believed there was an unseen substance, even in space, that allowed light waves to travel through the universe (Roy, 2017). To name just a few more, false, "scientific truths": 1. The Earth is not the center of the universe. 2. Our eyes do not see by sending out rays. 3. Margarine is not healthy for your vascular system than butter. 4. DNA is not three intertwined strands; it's a double helix. 5. The universe is not in a steady state, it's expanding. These are just five examples of many, real, ounce-accepted-as-truth scientific conclusions, that were in fact, wrong. This illustrates the emotional confirmation of truth that is based on false information. Being wrong, without knowledge of your false, "scientific" thinking, is exactly the same as if you are right. Everyone believing, especially in fields of science, one idea to be truth, exerts knowledge-limiting pressure on other scientists. If you were a scientist and said the Earth was not the center of the universe in the fifteenth century, you were considered an unlearned fool. So, new scientific discoveries are constantly debunking the truths of earlier scientific studies. There are multitudes of incidents where scientists overwhelmingly believe one "truth" for a time, then, this scientifically based belief is proven false. Also, a scientific theory can be true, and at the same moment, it could be believed by no one. Empirical science is intertwined with the previously mentioned, often-maligned, less precise way to discern truth… emotional reaction. Science is an important and effective way to discern truth, but it is not perfect. This is due to the fact that the very areas of science we humans' endeavor in, the hypotheses scientists choose to undertake, are guided by our emotional reactions from family upbringing, culture, and emotional investment in some belief. Scientifically, it is possible, we could be answering non-impactful questions. Truth is a reliable, predictable statement about some fact occurring in our world. What is a "true fact" in our perception may not be a true fact, due to differences in perception, current knowledge and cultural influence.

People, or groups, who make rules with their 'known truth' has the concomitant effect of making themselves superior. This superiority is seen in scientific circles as well, but is especially evident in large religious groups. There are countless examples of fictional, non-negotiable mandates declared by religious leaders. "If you don't *do* this, or *believe* this, you are not doing it right. "You are not following the truth." In reality, it's only ***their*** chosen truth. Even on a one to one, friend basis, the person who decides on any commanded rule, puts themselves (in their own mind) in a superior position. The friend, fellow group member, or fellow scientist, who doesn't go along with current thinking, is considered the wayward person. That pioneering person may be relegated to the position of being the "unlearned" person.

The same group pressure among scientists to continue to believe in some "truth," occurs exactly the same way in the realm of faith. Peer pressure to stay with what is currently believed to be true is powerful. There's a certain power in "knowing the truth." An example of a power grab from in the realm of faith, would be baptism. I could pick from literally thousands of made-up-I'm-superior-to-you rules, but this simple dip in the water will do nicely. Pre-Christian Egyptians, Babylonians, and Greeks, all have their ideas about baptism, long before Christianity came along. Christianity adopted the rite of baptism. (Baptism, 1881; Håland, 2009). Baptism has been thought of as a way to purify, to take blemishes away, to wash away sin, but that's just it... If you do not know the *true* meaning, well, then, you, to some people, would be considered inferior, or wrong. Is baptism a sprinkling of water? Is the rite of baptism full immersion into water? Does this rite of baptism have a physical bearing on a person? Or not? Should babies be baptized? Or older people who are able to make a logical decision? No matter what your beliefs are on this subject, when a person makes a rule and gets a following, those are the people who portray to others that they are "in the know," they "have the truth." Then followers have a belief, have faith, that their view of baptism, (or anything for that matter) is the ultimate truth. Never under estimate the power of groups of people believing in something that is not true. In the realm of faith, and the realm of science, large groups of people believing in something is a highly controlling entity.

People learning, coming to their own conclusions, for this book, the most effective method to come to "truth," is through three different approaches: reasoning, science, and your emotional steering. Just on an analytical level, it seems more effective to hear others' ideas and opinions, and understand why people think the way they do, then measure it against current knowledge, reason and science. Why do so many people just take someone else's word because *they say* they have a direct line to some divine source? Logically, that's just ridiculous, and academically lazy, but many do just this. The same stupidity of giving your life beliefs and time/actions to a government, is the same as giving it to a religious institution. To further illustrate how many people blindly follow, let's focus on the nonsensical faith practice done in the not too recent past... indulgences.

Throughout the centuries many Christians believed their sins could be forgiven by giving money to a church. This is the idea of indulgences, if a person pays enough money, they can buy forgiveness of sins (Shaffern, 1998). Logically, this is foolish on so many levels. Does God need money? Money, or any gem, or a mineral, or any other object which *people* decide has value, these things are obviously not needed by any almighty creator. Coins throughout history have been made of bronze, wood, copper, brass, gold, silver etc. God didn't decide these had value, a person did, and then others went along. People believed this nonsense because religious and civic leaders told them to. Indulgences is just one of many beliefs that most people currently think is foolish, but made perfect sense to people from other cultures in the not-too-distant past. Why give your beliefs and concomitant life actions to someone else, and what they think? It is more effective to listen to other people's ideas, but use your own experiences, your own logic, your own scientific research, to determine what is valuable. Using this approach will help us to come closer to what is truly valid.

A Muslim, a Christian, a witch, a Jew, a Buddhist, a Hindu, or a whatever, will tell you their scripture, or teachings, are the truth. There is no proof. All of these are stories, all of them are mythologies. The truths of these scriptures, at times, contradicts the other scriptures' truths. Furthermore, within each one of these major religions, individual people stand by, or break, the rules taught by their group, depending on their own varying life situations. It seems to be each person's duty to find loopholes in the "terms and conditions" of whatever local religion has been taught to them. How can a heaven or hell exist when right and wrong are so moveable? When good and bad is in the eye of the beholder? When teachings are in opposition, which person, or religious groups' teaching about what is right and wrong, *is the correct* right and wrong? People choose competitive domination; others pick helpful teamwork. People chose to live with an idea that they are superior to others; others choose to live humbly. People choose to acquire/hoard, others choose to be philanthropic. People choose monogamy, others choose to have many lovers. Who is living by the correct moral law? Which person? Which religion?

Pondering Our Existence for Practical Living

The moral laws by which we choose to live are often conflicting with others, and even our own normal choices. Take, for example, the founder of the Mormon church. Joseph Smith, in his late thirties thought he was right, good, and moral to get married to young teenage girls: Fanny Alger, Lucy Walker, and Helen Mar Kimball (Dias, 2014; Krakauer, 2003). Most people today, would consider these marriages to be wrong and pedophilic. Is my judgment about what is right and wrong better or worse than the millions of people who follow Joseph Smith's religion? There are many other people on this planet who would disagree with my view of Joseph Smith's actions. They may decide that taking young, and many wives, may be viewed as necessary to carry on their religion, and it is a wonderful idea. We all make decisions about right and wrong. Every person's choices are made because of their idea of what was right, or good, to perform, according to their own determination. Many who lead small and/or huge religious congregations, would say the good choices are good because the people in his/her congregation are following the things with which they agree. What is "right" is what the top leadership group states as being right and/or godly. People make ridiculous rules to please fabricated gods.

There is a similar parallel for religious leadership as there is for countries' governments, what they say is right, is what goes. If many people decide a certain spiritual teaching, or governing rule is correct, there is power in that belief. Most people just accept what their parents believe, or take on some other person's religious beliefs. These beliefs carry a lot of power and determine how people behave. Hopefully, the presentation in this book is different, because multiple possibilities are presented, and the ideas are backed up with science and logic. In the end, we individually choose what we think is truth. I saw on a social media site a "ground-breaking" article by one of the big religions. The religious authority/author stated that suicide is not a sin. Huh, interesting, well, I am impressed that he came up with this epiphany from heaven, but why should I care what this other person thinks? I'm not saying here whether I think suicide is right or wrong, what I am trying to point out is, why would anyone believe, one way or the other, because someone else told them to? This is so odd, yet this seems to be the way the vast majority of people make decisions. "Religious/government leader says it, so I believe it!" What we think of this issue of suicide, and every other moral issue, is profoundly warped through the lens of powerful religious organizations and/or governmental rule.

I do not have intentions to deride anyone's faith. In some ways, having blind faith in a mystical, all-knowing, Santa Claus/God figure, or a healing act (i.e. baptism) *is* superior to realism. A faith based in zero evidence, which most religions are, has benefits. For some people, there may actually be greater comfort in blind faith, in a mystical force of God or Gods. When a person can cast all their worry, guilt and/or anger away to an outside force, a force that will one day correct everything for you, that is an enticing covenant! To have everything be "made right," who wouldn't love that? It's a perfect way to escape the weight of life choices and ownership of reality. We acquire knowledge and settle on what is truth in different ways. Some people believe knowledge and truth is gained by faithfully agreeing to someone else's ideas. Many gravitate to faith as opposed to continual analyzing, objectively comparing, and continually searching. It is much easier to just accept other people's ideas. There are other benefits to blind faith, namely, one obvious benefit, a person fits in better with local culture. There is another obvious benefit for people who have complete and total faith in their decision of which teaching is truth, they have confidence. No more search needed, just deny new understanding, or fit in the new information (that may conflict) with their "rock solid" beliefs. People that have life all figured out, are usually confident and bold in their beliefs. They do not get uncomfortable when their beliefs are challenged, their self-imposed blinders give them an internal sense of superiority and confidence that they "know the truth." There is an easy confidence in any belief that does not have to have any proof. One can say, "I just know it's true." The veracity of these beliefs does not matter one bit, but the security in the mind is a huge benefit. These are impactful benefits to organized religion: confidence, safe belief, a path, traditions, culture, stories in art, holidays, and societal unification.

Hearing and studying other people's ideas, or listening to a guru of some sort, it is possible to come to a reason for living. This is one possible way to decide on the most beneficial method to live this life. However, it is more effective to listen to these ideas in the light of our actual life experiences, and have some basis in scientific and logical truth. We should use our best, latest, research in the realm of science. It is only possible to change for the better if theories about life are vetted through reason and science, *plus*, other peoples' interpretation. To just take, to just believe, automatically, something is true because of a religious guru, a political leader, or even what scientists say, will not get us closer to objective reality. There's an interesting book, *Soldaten*, that focuses on the main reasons Nazi soldiers carried out the orders to kill. It turns out, they were merely following the orders of their leaders (Neitzel & Welzer, 2012). We should never throw out experts, gurus, and leaders' ideas, but everything just needs to be further analyzed and rationally weighed. It is not beneficial to destroy others' icons. It is selfish and destructive to wreck others' firmly held beliefs, no matter how false they may be, including the art and artifacts that bolster their faith. Why destroy the beautiful artwork of Greco-Roman Gods? Or Buddha statues? Or Muslim geometric designs, Or Hindu deities? Or any religion? We should get enjoyment from this on an aesthetic level, learn from it, be entertained by it. There is no reason to be hostile to these likely false, yet entrenched influential mythologies. We should enjoy and appreciate the art, the stories, and the moral teachings, of any of the religions around the world.

Chapter 16

Ubiquitous Continuum #1: Good/Bad
Universal Pull of Right/Wrong with Afterlife Considerations

Right and wrong exists because we create these categories in our mind. We make judgments and act the way we do, because of our individual discernment of right and wrong. This idea of right and wrong is also directly linked to our benefit from any action, and the benefit, or detriment, that others incur from our actions. In addition to the fact of changeable right and wrong, every one of us, care about our actions. We make decisions every day on how to act according to our understanding of right and wrong. We could substitute "right" with "most beneficial path." We listen to that internal, emotional pull of what we think, and feel, is best. Actions are often thought of as good, or right, if there is a perceptible benefit to ourselves or others. Actions are considered good, if the level of selfishness is low enough so as not to hurt others. Good and bad are interwoven and moveable. Individual people's ideas, and society's ideas, of right and wrong, change over time. In fact, each individual's viewpoint of what is right or wrong is only true to themselves. Every human has their own personal view of what justice would look like if they had the omnipotent power of a god. For example, it is my view that whatever creative entity that made this realm, should stop the killing of innocent people. See? I have, just like everyone else, have a longing for what I consider to be my form of justice. As taught by many religions, a superior, and all-knowing God, would stop the unjust treatment, or murder of innocent people. Yet, this does not happen in this realm. Furthermore, there are many people throughout history who do not believe (as I have just put forth) that innocent children should not be murdered. A few examples: Nazi's murdering children, or the U.S. Calvary murdering Native American children, or the Jews destroying by the sword, every man, woman, child and all animals, in God ordained genocide as documented in the Old Testament (NIV 1984, Deuteronomy 20:16-17, 1 Sam 15:3, Jos 6:21).

Look at "truth" with logical reasoning: "truth" is changeable in the realm of science (new experiments debunk old scientific "truths"), "truth" changes with individual perception, and "truth" is defined differently by different cultures, and familial upbringing. Our actions are based upon our individual idea of the "truth," and what is right and wrong. There is an obvious fallacy in the thoughts of many people, that our choices in behavior can have impact on any after-life existence. Since what is a good and/or right action (or thought) changes from person to person, and culture to culture. Which one would be the truly right, or good? Is stealing wrong if a poor, starving child breaks into the rich person's house to get a bite of the rich person's plentiful food storage? The rich, fat, full-of-food person has food rotting in his/her house, and the starving child steals some… Is that wrong? Will the starving child be tortured in hell for committing this act of survival? Will the rich person be punished for greed? This is one of a million real situations that have occurred, and continue to happen here on earth.

Pondering Our Existence for Practical Living

Here are other moral scenarios: Will the wife who kills her abusive husband be punished, or rewarded in the afterlife? Will the father who kills a man who molested his child be punished, or rewarded? People will answer these moral situations differently, according to their own personal "truth" about what is right, and what is wrong. When what is considered righteous is so mottled, how can a possible soul be punished, or rewarded, by knowledge of malleable righteous actions? The problem with consequences in the afterlife is that we do not have any proof, or way to determine, definitively, what is right or wrong, all the time, for all people universally. As observed from the few scenarios stated above, right and wrong is situational, and up to individual interpretation. There are individual ideas about the afterlife. These ideas include overarching heaven and hell scenarios, a holding area, a reckoning, and reincarnation. Who wouldn't want to have a soul that flies when our physical body ceases? However, the reward or punishment in possible afterlives, of a possible soul, cannot possibly be based in the correct judgment of malleable truth and varied actions. Doing any given action is damnable in one culture, and admirable in another. Doing any given situational action is damnable to one individual, and admirable to another. This is not to say that right and wrong do not exist in the mind of each individual. Every human's choice of action already establishes that right and wrong exist individually. However, the changeable nature of right and wrong, in each circumstance, precludes a definite outcome in any dreamed-up justice in the afterlife.

Nagasaki and Hiroshima were great and wonderful triumphs… in the eyes of the United States government and military. The USA took revenge on a country that had bombed them at Pearl Harbor. On the other hand, to the people of Japan, to the people living in Nagasaki and Hiroshima, these "triumphs" in the viewpoint of Americans, are viewed as despicable acts of horror. Hitler's gassing the Jews to create a utopian society was a great and wonderful thing in Hitler's and his followers' eyes. If you read transcripts of Hitler speaking, he often states how God is on his side (Heep, 2020; Armbrüster, 2005 p 64). To the Jews (and the allied forces) Hitler was an evil, despicable, wicked person. Just these two, of hundreds of similar situations clearly show the veracity of what I'm saying, right and wrong is moveable. Additionally, this same phenomenon occurs in our everyday lives on a lesser scale. Getting fired from a job, hiring someone for a job, breaking up with, or cheating on your spouse, staying with your partner, getting in a fight, not getting in a fight… whatever… only *your* point of view determines your own choices as either right or wrong.

Just watch any movie, or read any story, and you'll find yourself supporting, or siding with one character or another. Oddly enough, humans have a great propensity to support and cheer for the underdog. Most people feel joy when a warranted revenge is completed in a story, an underdog wins, or an injustice is corrected. This is that same inner voice that guides our actions in real life. Listening to our inner voice, of what we think is best, should not be dismissed. This is why guilt is harmful and unproductive, and often unwarranted. Feeling guilty about past decisions, even if these decisions contradict so-called divine scripture, or the advice of a close friend, is not healthy. Yes, we should learn from our mistakes, but if you know what is right, for you, in that moment, you make a choice, do it, and then, do not feel guilty about it. Naturally, most people rationally pick a decision based on some perceived beneficial aspect, or personal reason. Then, later, they second guess themselves, and feel remorse for some decision that was the right decision in that moment. If one truly remembers the situation in context, from information at the time of decision, they would probably pick the same course of action.

We think one thing; we have an opinion… and we often change. Changing our beliefs when new knowledge is gained, is not an easy thing to do. This is especially true if everyone in your surrounding family/culture is not of the same mindset. Take for example the Salem Witch Trials. In the United States today, where I live, witches are funny. They're cute characters in movies. No one cares if you come right out and say, "I'm a witch!" In fact, I bet most people today would carelessly reply, "That's cool." On the other hand, back in Salem, Massachusetts in 1692, it is quite a dangerous thing to be accused of being a witch. Judges convicted many, and had twenty people executed for being witches. Sometimes, local societal pressure is blinding, unbearable and dangerous. The pervasive teachings at that time were that witches were real, they were in league with the devil, and it didn't matter one iota that the accused said they were innocent. Our indoctrination matters; our local cultural beliefs are powerful. What we are taught by the culture around us reinforces 'blinders' on our viewpoint. To the point where innocent people are murdered, this was certainly the case in Salem, Massachusetts in the late 17th century (Duntley, 2005).

Another example would be gay marriage. When I grew up in the 1970's and the 1980's, I did not know one single person that was openly homosexual. If they were gay, they didn't tell anyone. There were no homosexual people portrayed on the television, or in movies. Society in the United States, completely, throughout my life in the 1970's and 80's, taught that marriage was between a man and a woman. Politicians would often state this publicly to show they were "pro family." The idea of homosexuality being right/wrong has changed. Mirroring societal change, both power-house politicians Hillary Clinton, and Barack Obama, adamantly stated marriage was *only* to be between a man and a woman, but later in their careers changed their view (Sherman, 2015; Keith, 2016). From 1952 until 1973, The American Psychiatric Association stated that being homosexual was a mental illness (Anderson & Holland, 2015). It is now, not considered a mental illness. Here we are, some fifty years later, most people believe it is fine, and good, to marry someone of the same gender. Right and wrong not only changes with time, but also from culture to culture. Christians and Muslims would still say LGBTQ people turn to this "depravity" or "sin" because of a need for affection and the need to satisfy the physical body, it's an impulse that is considered sinful (Dialmy, 2010). There are those that have the view that homosexuality is a sinful loss of self-control. The truth is, the homosexual person has used his/her free agency to choose an action that is pleasing *to them*. Certain segments of societies around the world, have decided this is a naughty/wrong/sinful act, others say homosexuality is fine. Broadening out from the single subject of homosexuality, it is interesting how certain uncontrolled impulses are acceptable because society has decided so. For example: scripture reading, cell phones, eating chocolate, playing on the computer, scrolling on a phone, reading, are all prevalent addictions, yet these impulses, in my current society (by and large) are considered acceptable (Bierschbach, 2012; Condon, 2000).

Our individual labeling of thoughts and actions as being good or bad, right or wrong, means something. Think about the ramifications of this simple yet profound truth… we choose what we think is *a best* course of action. The actions we choose to perform, in every moment, whether they be destructive or constructive, are all judged, chosen by us, reasoned by us, and completed by us, because these actions "feel right" to us at the moment we act. This inner voice is *real, visible, experiential proof* for something beyond this life. Let the magnitude of this sink in, yes, the actions that constitute what is right and wrong is determined by each individual, and vary quite widely, however, the fact that *we all* have these visceral, emotional reactions to all events and occurrences means something. This is a clear indicator that actions matter. If there is no creative entity, if we are just a random collection of cells that ultimately have no reason to exist, why would anyone, ever, care about what is truth, or "right." Even the most avowed atheist is invested in telling everyone why there is no god. Without reason, or purpose for existence, why would the atheist **care** to convince anyone, of any "truth," especially about god's non-existence? The fact that we are **all** concerned/worried with what is right and wrong is proof, in and of itself, there is some reason for our concern. Why would we care about any action or thought unless there is some reason for this concern of what is right? Think of this situation conversely, if our actions *didn't* matter, we humans would be apathetic, we would not feel any urgings of right/wrong one way or the other, but this is certainly **not** the case. In fact, the opposite is true. Passing laws, killing others, arguing with others, judging others, punishments and rewards, are all completed by humans because of what we each individually, and collectively, believe to be right or wrong. This is strong proof that there is a creative entity, and reason for our existence. If there is no reason for our existence, our concern for right and wrong (which guide our decisions) would not matter; we would not care one iota about our own actions, or the actions of others. Some people may SAY, with their words, "I don't care." However, this is not bore out in their observable activities, and visible, emotionally-charged actions and words. There are definitely different levels of care about actions between people, but in the end, we are all concerned with our words, decisions and actions.

Even though what is deemed as right and wrong is constantly changing, and good and bad are completely interwoven, there most certainly exists an internal moral compass for our actions. There unquestionably exists a right and a wrong in the emotion of individuals and/or local culture on the whole. Even if you believe our conscience came by evolution, there is undeniably an inner voice, an emotional guidance by which we all make decisions. Why would we have this "right and wrong" idea within us? There must be some reason for this awareness of right and wrong within ourselves. There must be some effect, otherwise, humans would not care about any decision. Some believe this conscience could be a naturally evolved survival aspect of our species. They say that our decisions we make are biologically, and automatically chosen, due to humans' evolutionary survival while living in close proximity to each other (Tomasello, 2013; Gillett & Franz, 2016). However, there is strong evidence to oppose the view that right/wrong judgment came from evolutionary means, because moral decisions can lead to our own biological death. The consequences of moral action are often ***not*** evolutionarily advantageous. Morals are standards of behavior, chosen by each individual according to what they feel is right or wrong. Since these moral values change and vary, this precludes any evolutionary benefit. Also, if we could imagine a human without the emotional pull of morals, that human would be *vastly* more efficient in taking action. A human unencumbered by feelings of worry, feelings of guilt, feelings of shame, due to moral conviction, would take whatever action is needed for survival and reproduction without concern. Here are just a few of many possible examples where our inner moral voice could get a person killed. Say a person picks the moral of courage to defend a young abused person. This decision may lead to actions which get them killed, and ends their chance of passing on their DNA. Or, the moral of an adulterer being put to death. The fact is, the inner, moral voice of right and wrong can often times be destructive to ourselves, and others, both physically, and socially. So, there is little evolutionary advantage to our varying moral judgment of right and wrong. In reality, every right/wrong moral under the survival-of-the-fittest umbrella (For example, the choice of when to be violent, aggressive, conniving, etc.) could be disadvantageous to passing on one's own (or stopping others') DNA. The morals of competing for resources, and competition in mating, definitely, *do not* foster morals of social coexistence. Individuals within group living situations compete for resources and reproductive success, even though the moral of egalitarianism offsets destructive standards (Gavrilets, 2012).

Another reason morals could not have come about by evolution, is because half of all morals are unfavorable to communal living. Selfishness, and personal gain, are not applauded morals, but these chosen morals are prevalent. Some of the *virtuous* morals would include: honesty, respect, compassion, empathy, kindness, generosity. Unlike outward actions, virtues (the constructive morals) seem to be considered beneficial across time and culture. However, there are just as many (opposites to the virtues) morals that are destructive to living socially: scheming, conniving, aggression, domination, selfish gain by disrespecting others, brutality, superiority/imperialism, and miserly hoarding. It is infrequent for people to understand that "good" or "bad" morals are standards chosen by which to live, each person's decisions (on a constant basis) are a result of each person's idea of right and wrong. Virtues are the *positive* morals that do help, the virtues build, the virtues bolster communal living. Conversely, and importantly, there are just as many morals (that are not virtues) that destroy, as there are morals that build. A person would have to only acknowledge the morals that benefit communal living, and ignore destructive morals, for morals to have any evolutionary benefit. Since there are morals that are absolutely destructive to our species, and world, it is not logically possible that morals exist due to evolution and/or safer communal living. To make this point very clear, there are morals that are taught (indoctrinated) that are severely destructive to survival, and passing on one's DNA. Think of the millions of people who have been murdered throughout history due to the *moral teaching* that "my particular religious doctrine is superior to another." Or, that one form of governmental leadership is more advanced than another country. These morals are responsible for many wars, and untold numbers of deaths. These are evidences that moral, right/wrong decisions, came about from some aspect other than evolution, more likely, morals are an imprint of a creative entity.

How other animals' morality works, as with everything, is debatable. Some think morals are biologically innate (Hauser, 2006). It is evident, we can easily observe, some understanding of right/wrong in certain animals. Humans self-inspect, and judge our own actions. We humans make choices and act how we want to, on a constant basis. Possibly, other animals do too. The indication would be our own pets. I have seen my own dog certainly have a guilty face expression when she does something disobedient. At this point, it is not possible to know what level of cognitive functioning is needed to feel the pull of right and wrong.

Longing for justice, and making things correct, may mean different things to different people, but **everyone** wants fairness and justice in their own view. What is right? In many cases, what is right for one person, is not right for another. And "fair" for one person means being "unfair" to another. Every human has a version of right and wrong and justice, this is probably some sort of imprint from our creator(s). Since right/wrong urgings often harm our evolutionary ability survive, and reproduce, right/wrong is more likely an indicator of the intellect of our creator(s), because *everyone* has this same longing for justice. The lack of miraculous, divine intervention has no bearing on our longing for justice, or what we think is "right." Even though what is considered just, or fair, is changeable from person to person, still, everyone longs for what they think is right, just, and fair. It is observable evidence – we all want fairness, and our version of justice.

What is considered good or bad has impact on fitting in with local society. Doing what others think of as "popularly correct," can actually put lives in danger. In order to fit in, propaganda sways both the learned and the unlearned masses. Persuasive, life altering information, in the form of societal pressure through personal relationships, advertisements, song lyrics, movies, books, and internet searches, this exposure directs our lives. Humans will latch onto anything that is familiar, and/or foisted upon us as accepted as the current, trendy, correct, best way; comfort and acceptance is a powerful deterrent to knowledge, truth, and real freedom. Any person who dared to challenge the judges during the times of the previously mentioned Salem Witch Trials, was putting themselves in absolute danger. Along the same thought line, a common, everyday situation is which clothing we choose to wear. Also, what hair style, or whether to have a tattoo. These fashion choices may cause a person to be either ostracized, or accepted. Death due to fashion choice isn't currently a normal outcome in my current culture, but the danger of wearing "out of style" clothing, or having a tattoo, will, at the very least, cause people to be treated one way or another. What is good or bad in the perception of physical beauty, and stylistic fashion choices, is idiosyncratic. However, for better or for worse, this perception has a powerful effect on the treatment each of us receives.

What is good, or what is evil, is idiosyncratic. Aside from a silly, fictional, devil character influencing people, there are some real reasons for what looks like unhealthy/evil/bad behavior, chosen by individuals in varying situations. These could include heredity, absence of good parental instruction, abuse or neglect, or prevalent destructive behavior on a local/cultural level. All of these reasons for detrimental behavior have nothing to do with the devil or demons. All of these aspects could severely limit a person (especially in childhood) from having a chance to choose actions that are seen as good, or beneficial, in their current culture. Children raised in any environment (evil or not in the sight of others) will be influenced by his/her upbringing. Yet, people rarely, completely lose his/her ability to choose what he/she thinks is the most beneficial course of action.

Assuredly, many laws/rules based on morals exist because acting in certain ways does benefit society. What is considered "sin," or "wrong," are often actions that are unfavorable to existing close to other humans in a surrounding society. I previously covered how many morals do not benefit societal living, but the "good" morals, the virtues, certainly do help communal living. For example, stealing being considered wrong, in most cases, helps the entire community. Lately, in the United States, there are many publicized news stories of crowds of people rushing into stores and stealing everything they can carry out (Belgum, 2023). One documentation of a mass robbery was in a gas station/convenience store. The group of about thirty young people came in and just opened packages, ate food, stole money and destroyed the store by throwing merchandise all over the place (Moran, 2022). This is happening in many U.S. cities where the people, and law enforcement, do not stop the thieving vandals. Sometimes, just one or two people will brazenly, confidently, walk into a store, take whatever they want, and walk out. The law enforcement has recently become weak (to non-existent) in many cities of the United States. There are many problems with not following the societal rule, "the sin," of stealing. First, in many of the incidences, the people who are working in the store are disrespected and physically hurt. Second, all the suppliers, the people who transport these goods, the owner of the store, the employees of the store, all these people, their money that they are earning by working, is stolen from them. The selfishness and greed of the thieves is clearly hurting other people, even if it is beneficial to the robbers. When mass theft occurs, often to bigger retail stores, or just one mass theft from a small store, the stores must close down. Which brings us to another detrimental ripple-effect of stealing, all the people close by, that shopped at that store, no longer have that place to procure what they need. Lastly, if this rampant stealing occurs, it will destroy the work ethic of all the people involved in the sale of goods. Why would anyone keep working in a legitimate manner when they can just steal someone else's goods? Actions are often considered "evil" and/or "wrong" because they injure communal living.

This is why the institutions of government, religion, clubs, and groups exist. Any group gathering exists, because people of similar values meet to bolster, encourage, or even to enforce (as in the case of laws and government) their beliefs. Religion is an effort to bend as many people as possible to believe, and live their lives, according to someone else's idea of what is beneficial and the "correct" way to live. Everyone chooses what actions he/she thinks are good, or virtuous, but most people are swayed by the massive pressure of predominant groups. The morals that are considered virtuous lead to the creation of laws and rules, and help in living in proximity to each other. Oppositional, moral choices (those that are not considered virtuous) made in the moment, all arise from what is beneficial for selfish reasons such as, arrogant superiority, indoctrinated beliefs, competition, and individual survival.

Chapter 17

The Violence Compassion Continuum

If actions that are detrimental to society (the opposites to virtues) are predominantly utilized, that society will not flourish. Let's focus on practical outcomes of mercy and violence. Would not humans on the planet Earth be better off if they did not have to worry about another person raping, beating, killing, or otherwise physically hurting them? In an environment of excessive fear and danger, this lowers the quality of life. In such a violent setting, on the average, their death rate would be at a younger age than those in a society that follows the morals that are beneficial to communal living (Kan et al., 2021; Holaday et al., 2021). Society not murdering, not stealing, not physically hurting each other, is safer and more productive than a society filled with danger (Aburto & Beltrán-Sánchez, 2019). Letting others have rights to do with what they will for their own body seems like a very basic right. A safe environment is an easy way to improve all of society. What is considered right and wrong varies and changes, however, kindness just seems to be more beneficial than aggression for living near one another. A person in a safe environment can flourish, a person living under threat, is troubled. If people in a society chose aggression and violence all the time, the society would not function, and would eventually cease to exist at all (Aburto & Beltrán-Sánchez, 2019).

I saw on the news, people of the Muslim faith mercilessly beating and torturing prisoners that were bound by ropes (Westcott, 2016). It was difficult to watch, the butts of the aggressors' guns, whips, and fists were all lambasting these bound captives. There have also been broadcasted beheadings, and setting people on fire, because the person did not believe the same way as others do. This is just one example of unwarranted brutality that is replicated around our world every day. Wouldn't it be wonderful if the guy giving the beatings, or about to light the fire, or pull the trigger… wouldn't it be great if he just put the torture instrument down? What if he just set the person to be tortured free? Why do some people, rational human beings, choose mercy, and others choose merciless violence? Of course, some people would disagree with the judgment that mercy is preferable to wanton violence. Again, we have this natural and automatic longing for our idea of what we feel is justice. This feeling, involving the balancing of the scales of justice, by way of violence or compassion, could be a product of human interaction and societal norms. In any case, the feelings are certainly present. It is just as plausible this longing for justice, the correct measure of mercy, or punishment, originate from the imprint of whatever created us.

The U.S. Calvary, massacred innocent women and children, even infants, at Sand Creek, Colorado in 1864. Gold was found in Colorado in 1858, so the USA decided to further restrict where Native Americans were allowed to live. What if the soldiers said, "No, this is not right," and just rode away? Their conscience didn't tell them to ride away, or maybe it did, but cultural pressure overrode inner feelings of right and wrong. It's quite possible the soldiers' upbringing and indoctrination allowed them to feel no wrongdoing in the murder of these defenseless Native Americans. Here is an actual, proudly-racist quote from John Milton Chivington. Chivington was a colonel in the U.S. Calvary that led the massacre at Sand Creek, "I have come to kill Indians, and believe it is right and honorable to use any means under God's heaven to kill Indians… Kill and scalp all, big and little; nits make lice." (Brown, 2001). The indoctrination of our upbringing, or cultural pressure is powerful. The indoctrinated moral chosen, at times, is lethal brutalism. This is the same true belief and vigor for death that the Nazis had, the Portuguese slavers, the Middle Eastern ISIS members, the list could go on forever. Many people would prioritize mercy and peace much higher than Chivington did in 1864 at Sand Creek. Young, native girls, pregnant women, the old people, and juvenile native males… all massacred. The troops willingly, aggressively, smashed infants' brains upon rocks (Jackson, 1994). Who can say one person's (or culture's) prioritization of values is better than the values Colonel Milton Chivington, and the U.S. military had in 1864? I can, and you can, because each individual knows what is right and wrong from within our individual selves. My personal idea of right and wrong is vastly different than the choices of the soldiers at Sand Creek, but still, however different, the idea of right and wrong is always present. Harkening back to my earlier point, incidents like Sand Creek, Colorado, would be a *great* time for a miraculous intervention from our designer(s) if there was any intervention at all.

Chapter 18

Afterlife Ideas, Faith Shifts Blame

Does your individual soul's torture or contentment depend on choices, thoughts, and actions that we make throughout life? Earlier in this book, I put forth evidence of brain waves outside, separate from our physical body, possibly a soul. These study-able evidences include: 1. Mind cloud 2. Shared, waking reality 3. Time non-existent for a person under anesthesia, yet they still exist 4. Near death experiences 5. Brain waves outside the human brain (able to be read and drive a car for physically disabled) 6. Scopaesthesia, just to name a few previously covered. If a part of our makeup is a non-physical soul/spirit, this part of us could continue after our physical body ceases functioning. Ideas about what happens when we die, do have real impact on our everyday decisions. Let's weigh all options, focus on the observable, and then individually, we may determine what we think is the most likely possibility.

No one knows what happens when we die. However, it is possible to look at proposed options and dissect these ideas. There are organized religions with all kinds of interesting ideas. Heavenly afterlife, an afterlife that usually just a man's paradise. For example, an afterlife full of beautiful virgins. Or, an eternal warriors' feast with ale running continually. According to some, you may even get to rule your own planet in the afterlife. There are major complications with all kinds of heavens, hells, or other afterlife scenarios that people invent. No afterlife concoction exists with any definitive proof. Heavens and hells with virgins, fire, worlds to rule, unlimited procreation, imbibing alcohol, rest, peace, fighting, work, bright-light, family reunification, judging worlds, singing, torture, trash dumps... these are all real fabrications by people of our past and present. Ostensibly, to inform us what our experience will be in the afterlife. Many people truly believe these cockamamie (albeit possible) afterlife existences. These scenarios of the afterlife are all possible, but unprovable. There is no basis in anything other than the imaginations of people, so all these versions of heaven are only possible through hope and belief.

Since there are so many varying views of the afterlife, obviously, no one knows what really happens. Regardless of afterlife possibilities, should we not at least try to work toward a fulfilling existence here on earth? Why wait? This is a problem with placing belief in a corrective afterlife, this belief can cause apathy in doing something now, to correct perceived injustice in *this* life. We should concentrate on making this life as great as possible. Why take limited, or no action at all, because there *might* be a reckoning in the afterlife? As limited as our senses are, we have to trust our senses in the here and now. I remember watching the O.J. Simpson trial on television. This famous court case involved a National Football League athlete, who was on trial for murdering his wife (Stephanopoulos, 2024). One of the guys I was watching the trial with, said something like, "Who cares if O.J. murdered his wife, he has just a few years left on this planet, and he will die and have answer to God." This is the danger, we stop enacting beneficial and/or proper decisions, big or small, personal or societal, because some afterlife entity will straighten everything out. Many delay, or never do anything, because they are relying on an afterlife reckoning. The reality we perceive is now, this moment, this is precisely where beneficial decisions need to be made.

Relying on a reckoning in the afterlife, affects decision making in this life. It is a way of shifting blame and responsibility from self to a being outside your control. Many people attempt to separate their own problem(s) to a mythical third party, a god, or gods, or the devil. Many people choose to "pass the buck" on any of their own chosen actions. These people might say something similar to, "The devil made me do it," or "an evil spirit led me to destruction," or some other, outside, evil force made them do such and such. When you think about a devil character, logically, he just seems silly, juvenile, and truly comedic. The devil has the duty to make life difficult on us? To make us scared? The devil makes people do bad things to other people? Many of the big historical religions of the world have a demonic figure that is in opposition to what is good in their teachings. Is it so hard to believe *you* are in control of *your own* actions? As opposed to some devil character, an outside personal force to blame? The devil is just another excuse; similar to God/Holy Spirit, the devil is a way of "passing the buck" for bad behavior. No, the devil is not in control of your "bad" actions, you are.

In the same way, no God, or Holy Spirit is in control of your "good" actions. This blame of personal thought and action may not be effective, or true, but this shift of blame definitely takes the pressure off humans for their individual responsibility. When a person says they are, "carried along by the Spirit to do good." A person gives themselves an easy out, they can say, "All the thoughts and actions I did wrong will be wiped away by my caring, heavenly-father figure, the Holy Spirit was guiding me." This shift of responsibility from self, to outside God(s) not only affects actions taken, but also installs their personal view of their God's idea of right and wrong for all people. Stating that if people don't follow what "He" (their own God) sees as true, they "will be taken by the devil," or "go to hell," or whatever consequential afterlife in which they choose to believe. The consequence of myopic, obedient behavior is often said to be rewarded in heaven. Just as with there being no provable afterlife scenarios, the reality is, there is no way to prove the devil, or concerned God(s) exists. Conversely, instead of outsourcing responsibility, impactful and superior decisions will be made if a person takes responsibility for their own thoughts and actions. Shifting blame (for what you did) to demons or gods, or Satan or Jesus, or installing one personal view as the one true god for all, is not at all based in logic. Impactful and superior decisions will be made if a person takes responsibility for their own thoughts and actions, not some product of an outside god source.

Chapter 19

Afterlife Ideas: Reincarnation

Many people believe in reincarnation. It is the belief that when you die, you come back to earth either as a human, another animal, or even a plant. Notably, the massive religions of Buddhism and Hinduism, as well as many individuals have this belief. The soul (called the anattā in Buddhism and ātman in Hinduism) is the inner most essence of oneself. It is the unchanging, independent, core of a person (Atman, 2018). When a person dies, they go to another world, a second world, where the thoughts and actions of their previous life are judged. Then, they come back to Earth as either a higher level being (higher lifeform, with higher level of consciousness) or a lower conscious lifeform, such as a plant. People of the Hindu faith believe a god (Isvara or Shani) judge your good, or bad deeds, and decide to what being a person will reincarnate to upon their return to Earth. In Buddhism, they believe there is no god judging and making decisions. Rather, a person reincarnates according to their karma just naturally, and automatically, as a system already set in place. In both Hinduism and Buddhism, a soul can escape the continual reincarnation back to Earth, by living the right way. There are many overlapping, agreed upon "right" ways to live

(Tomkins et al., 2015; Cope, 2006; Baruah, 2019) according to both Buddhism and Hinduism:

- Detach from the material world/meditation
- Gain knowledge
- Selfless service/eliminate greed
- Overcome desires, not craving, detach from desires
- Living in a way not harmful to other beings

As with earlier critique of shifting blame to angels or devils, there are logical problems with the idea of reincarnation as well. Reincarnation is just another possible story to believe, as with all other non-provable stories, the karma/reincarnation cycle were made up by people to attempt to enlighten others in what happens when we die. A person in Buddhist/Hindu beliefs does take accountability for their free will choices, but a person must *agree with*, *and follow*, someone else's idea of what is good. Is not having desires, (Samudāya/craving which causes suffering) and not striving a good way to live? Maybe yes, maybe no. To reincarnate to a superior being when you return to Earth, these are the "correct" ways to live. Not striving, and not desiring, are two important tenets in both Hindu and Buddhist teachings (Tanaphong et al., 2019; Saltzman, 2017; History.com, 2023). What if a person does not agree that not striving, and not desiring, as being a good way to live your life? Some people would never live in the way of Buddhism or Hinduism, but rather strive to have all the experiences they can, make goals, and possibly even accrue material possessions. It is certainly possible that a person could believe that a great life, a wonderful way to live, would be to chase *all* their desires. Some people get a real thrill out of fulfilling their desires, not denying them. So, whether an afterlife judgment is by some god, (Hinduism) or just happens (Buddhism), what *true*, what *fair* assessment of karma fits which human/plant or animal a reincarnated soul will inhabit? This point may be easier to see with a starker example of how cows are treated. In India, cows are considered sacred by most people. Cows are intertwined with Krishna (central Hindu deity) and protected by law (Parikh & Miller, 2019). There are laws against slaughtering cows because they are on similar footing as a human. Plus, there is the overarching teaching of not bringing intentional harm (ahimsa) to another living being (Saxena, 2017). The non-guilty, not sorry about, no remorse person in the USA, Italy, Brazil (any non-Hindu/Buddhist area) *enjoying* a juicy steak will (unbeknownst to them) be destroying their chance at getting a better existence in their next reincarnation to Earth. Judgment on what is considered sinful behavior is so varied, who can guess the positive, or negative, impact on a soul after karma judgment when a person returns to Earth?

In addition to the possible differences and disagreements in what makes a worthy, "karmic, level-up life" in Hinduism or Buddhism, there is another problem with reincarnation. The soul of each being is inexorably connected with sins (or good, depending on the viewpoint) from previous incarnations here on planet Earth. Let's say we all agree with the good-karma-building idea of selfless service to others. Let's consider the person who lived life opposed to selfless service. They lived their life with greedy desires to achieve great wealth. When the greedy person dies, their soul would reincarnate to a lesser being. A person's "sins" are always baggage in subsequent lives. Not only does a person have to atone for the bad actions in this life, but the incredible pressure of previous lives weighing upon individuals as well (Burley, 2013).

While we are discussing afterlife scenarios, and the impact these ideas have on how we live our lives, it must be said that there is some evidence for reincarnation that must be examined. We have just examined the logical flaws of reincarnation, but Dr. Ian Stevenson, an American psychiatrist (1918-2007) and many others (Dr. Jim Tucker), have documented young children (thousands of cases) who remember details of a previous lives (Tucker, 2008). Incredibly, some of the toddlers talking about their previous lives have given verifiable facts, facts the little ones could not possibly have known. Equally improbable, researchers have documented hundreds of cases where birthmarks and birth defects on the living person who is recounting their old life, match the mortal wound of the deceased person being remembered (Tucker, 2005). How can this be? One, obvious explanation is reincarnation, the soul was reborn into a new person, who is now a toddler remembering their past life. A second explanation, it could be that the soul, or inner essence of a previously alive person, has housed inside the new person. Similar to a "possession." A third explanation, the young person remembering their past life is accessing some cloud/database which has souls in it. It's unprovable, but it must be considered, many people believe in an "Akashic Record." This is the name given to an immaterial collection of all thoughts, emotions, and events that have, or ever will occur, in this world. Whatever the reason, there is excellent documentation of toddlers remembering past lives (Stevenson, 1975; 1983; 1993; Levin, J. (2020).

Chapter 20

Ubiquitous Continuum #2: Constructive/Destructive Murderous Aggression & Caring Compassion

Every aspect of life can be put on a continuum. We only understand laziness, because we can compare it to industriousness. Kindness is only known, because we also know cruelty. How could we ever know love if we do not encounter hate? The opposite allows us to define anything, and everything, that we know. Let us focus on two basic, observable, oppositional forces in everyday life: destruction and construction. The earlier discussed good/bad continuum is separate from the construct/destruct continuum, because good/bad is mostly decided cerebrally. The construct/destruct continuum is based in our physical realm. Placing "might makes right," in juxtaposition to "unselfish / kindness" under this broad destructive versus constructive umbrella. This continuum of destructive/constructive is one of the broadest, most impactful of all the aspects of life. The constructive/destructive scale is present in all our actions and thoughts, and in all that happens in our surrounding environment. Just as the predictability of nature is evident, it is also clear that the creative entity of this realm wants both overarching actions to exist: building and destructing. The destructive lion tearing out the throat of the gazelle, exists alongside the human saving an injured bird. Is one animal right, and the other animal wrong? We see constructive and destructive actions in everyday life. Destructive-might-makes-right and constructive-unselfish-kindness are sometimes intertwined, and which action is which, is completely determined by individual perception, nevertheless, destructive and constructive are both present.

Opinions change about what is considered destructive or constructive. Plus, similar to what we decide is right or wrong, what is considered constructive or destructive, is completely in the eye of the beholder. I am sure that every person reading this can think of ideas they had when they were young, and how those ideas have changed over time. If a person doesn't change, there is a problem with honest self-examination and personal growth. Here is a personal example of change in thoughts of constructive/destructive behavior. When I was young, say, before 16 years of age, I would eat everything with zero guilt. Steak, shrimp, crabs, chicken, you name it, everything. My mom used to say that if my metabolism ever caught up with me, I'd weigh 400 lb. Somewhere along the end of high school, into my first years of college, I began to feel guilty about eating other animals. I don't remember an exact incident, or learning from anyone some specific teaching, but I attempted to go against the natural setup of this world, and I become a vegetarian. Deer, tortoises, rabbits, etc. are biologically vegetarian, but they still have to kill living vegetation to maintain their lives. If anyone honestly looks at life on this planet, in order to live, one **must** kill something else. Either a living plant, or a living animal, needs to be eaten for us humans to survive.

Pondering Our Existence for Practical Living

What we determine to be constructive or destructive changes even within our own thinking. For a time, I would not eat meat. When my own conscience was deciding on guilt for eating animals, oddly enough, I felt no remorse, or concern, for plants and/or animals that are ugly. Say, for example, mosquitoes, roaches, earwigs and spiders. I could always smash them with no guilt in my heart. Additionally, I had a clear conscience about murdering literally thousands of insects on the front of my car as I drove the roads. Eventually, I determined that to oppose my natural, omnivore, biological-self, was more difficult, (and not as beneficial) as following my temporary belief that I should not eat other animals. Nature, and the obvious truth (I salivated when I smelled juicy burgers on the grill) that humans are biologically, omnivores is powerful (Boehlke et al., 2015; Daujeard et al., 2020). This made me drop this culturally pressured, moral, vegetarian-stand that I was taking. Importantly, plants are living as well, there are no outcries for animals to stop eating plants.

There are recent studies proving, it is a fact, that plants are able to make airborne sounds, they "scream" and feel pain (Karabey, 2010; Mishra et al., 2016). It has been scientifically proven that plants have feelings, and reactions. Hertz denotes sound wave frequency, our human ears only perceive from 20 Hz up to 20,000 Hertz, but we have instruments that can hear/measure sound from plants in frequencies from 20,000 Hz up to 150,000 Hz. Ultrasonic sounds (Plants can 'scream' and react) are emitted by stressed plants (Mishra et al.; Robert, 2023; Berkowitz, 2024). These sounds produced by plants can be detected several meters away from the actual plant. Sounds from plants have been recorded and classified. Scientists, solely by examining the recorded sound emissions from plants, can identify the condition of plants, including dehydration and/or injury (Khait et al.; Mohanta, 2018; Dicks, 2007). These informative sounds may also be detectable by other living organisms. One way or another, living animals and/or plants, need to be consumed by humans for our survival. A vegetarian cannot logically be viewed as having superior morals, they still must kill and consume living plants to survive.

Changing one's ideas on moral decisions with meat eating versus being a vegetarian, is just one example of changeable viewpoint on the ever-present constructive/destructive continuum. Many animal rights people are opposed, on moral grounds, to destroying animals to eat meat. If these people got into positions of power, they would mandate their ways upon everyone else. What about the person who derives extreme joy, and verifiable health benefits, (Zhubi-Bakija et al., 2021) from eating meat? What about the person who follows our human biological needs as omnivores? The perception of eating meat, being right or wrong, brings these meat-eating people into conflict with vegetarians. My current belief is that whatever food or drink goes in my mouth has no bearing on whether I am a good or bad person. I personally determined, eating meat is good, constructive, and healthy. On the other hand, what comes out of my mouth, the words… those can actually hurt/harm others. Right and wrong, and what is considered destructive, or constructive, changes in individual thought, and changes over time.

To further humanity, to progress to new and better ways of existing here on Earth, we have to analyze our own individual choices. Even writing about my changing ideas on being a vegetarian, this will probably aggravate and annoy both people who are strict vegetarians, and it will also annoy those of us who do eat meat. This is because we have all been taught in one way or another that one of these approaches is *right*. It is only possible to find out what makes you the best human being you can be, on an individual choice basis. We consider all thoughts, words and actions, we perform and choose what we think (and feel) is most beneficial. Notwithstanding the pressure of our family, local society and governmental laws.

Pondering Our Existence for Practical Living

We, humans, defile, we can't help it. It is important to note that other animals defile as well: magpies, ants, ticks, all these animals have a destructive element built in. I've seen magpies kill other song birds for no reason whatsoever. Termites and ants destroy wood. Ticks spread diseases when they bite. Orcas torture seals by slinging them into the air. Humans destroy other living things and alter our environment to survive and thrive. Humans manufacture massive amounts of waste. Of course, humans are destructive, but everything is destructive. Destruction, even though it is not thought of as desirable, or as adored as construction/building actions, it would be false to deny that destruction is imbedded in all aspects of this existence. Recall my earlier point on vegetarianism, all must kill, or annihilate some other living thing to keep our own lives. Aggressive, destructive actions are *built in* to this existence.

We took our children camping throughout their upbringing. Sometimes we pulled off the side of a forest road and we'd sleep in a tent for a night, or two, and hike around the forest. We brought a shovel on these mini trips, and one of the fun activities our kids enjoyed doing was to watch what we called "Ant Wars." There were many, large ant hills built by these industrious creatures. Each ant colony had its own two to three-foot-high mound. Interestingly, if you scoop a shovel full portion of one ant hill, with ants, and throw them on another ant mound, can you guess what happens? The ant hill that had the "foreign" ants thrown on top, makes all the ants immediately go into a frenzy. The two separate ant hill colonies start ripping each other apart. Literally, the ants rip off each other's' heads and limbs. Ignoring the fact that my family enjoyed these battles, it teaches us something very clearly. Why do the ants automatically go to war? Why doesn't the ant colony, when mixed with another ant colony, welcome their fellow ant brethren? The answer is obvious, that just naturally, there is an imbedded destructive element in our existence. This does not just concern ants, but spiders versus praying mantis, beetles versus crickets, cockroaches versus ants, the list could go on forever. We observe wanton, destructive behavior in humans *and* in all of nature.

On a chromosomal level, chimpanzees and bonobos are the closest animals to us in terms of DNA. Long term studies show chimpanzees having outbreaks of lethal (yes, fatal) violent attacks upon each other (Morrison et al., 2020; Pandit et al., 2016). These attacks gain the dominant chimpanzee group new food supply, new territory, and mating with females outside their own chimpanzee community. Roaming male chimp groups attack opponents whenever they have an advantage in numbers. Interestingly, the attacking chimpanzees don't stop when their victims attempt to flee. The attackers collectively pin down and maim and/or kill their victim. Chimpanzees engage in warfare whenever they can gain an easy victory. Chimpanzees will also invade a neighboring chimp community, seek out, and kill chimps from other groups. Conversely, within chimpanzee communities there is often cooperation: grooming, shared food, and lethal chimp-patrol-parties roam in a single file lines. However, even within a single chimp group there are fights, posturing to see who is the alpha male, and ranking after that. This destructive violence determines who will mate with female chimpanzees. Even in gorilla groups in the Republic of Congo, less dominant groups had smaller territories and avoided the dominant gorilla groups' areas. There is territorial based violence while simultaneously affiliating with their own group. There is some intragroup cooperation, but undeniably, fighting, aggression and battle is built-in (Morrison et al., 2020; Pandit et al., 2016).

Bonobos are very similar, genetically, to chimpanzees and humans. Bonobos are skinnier and have smaller heads than chimpanzees, they also live in Africa quite near to chimpanzees, but they are a different species. Bonobos, overall, are far less aggressive than chimpanzees. Yet, when separate communities meet there is still hostile posturing, and even physical attacks, by alpha male bonobos. Bonobos are different from chimp interactions because there has been much observation of different bonobos communities, sharing food, grooming each other, and having intercourse with the other group. Chimpanzees would never do this (Machanda, 2018; Moscovice et al., 2022; Pandit et al., 2016).

Even at the microscopic level we see microorganisms hunting, and warring with each other. The fact is, heterotrophic organisms (any living thing that must consume other organisms for energy/life) prove that in our world, aggression is built in. T-cells in our human body attack cancer cells. Neutrophil in our blood stream, finds bacteria and kills it. Dileptus is a single cell organism that has a long snout in which it uses to stun, and eat, other microorganisms. Lacrymaria Olor feeds on ciliates, flagellates, and amoebas. "Phagocytosis" is when a cell surrounds, ingests, and destroys other microorganisms. There are many, many more murderous microorganisms that could be listed. It is a definite, observable fact, that destructive aggression, and even annihilation for survival, is built into this earthly system (Mills, 2017; Cooney et al., 2024; Flaum, 2023; Miller, 1968).

We see proof in nature of murderous action at the microscopic level, in primates, ants, insects, sea life… everywhere. Humans act the same way. There are individual people, and yes, even entire nations, throughout history that have the propensity to dominate. This aggressive/imperialist thought process occurs on an individual basis as well as a group, or nation level. Sometimes individuals, nations/groups of people will annihilate another person, or group of people, unto physical death. People in bars, in schools, in the work place, in the home, can at any point become an aggressive destructor. The aggressor participates in forceful actions to make things go their chosen way, or simply just think of themselves as superior. It could be the aggressor is just selfishly wanting to take what the other people own. For selfish reasons, aggressors feel they have the right to destroy others. These people/nation groups are bent on destruction and domination.

    Both constructive and destructive behavior can feel good emotionally. On social media I see many real, street fights. Not fixed, not like fake boxing, or wrestling on TV; I am talking about *real* street fights, where people are angry and upset with the other person. They will punch each other in the face, scratch, pull hair, body slam, whatever they can do to win the fight. In many of these videos, when the more aggressive, stronger, or person with smarter fight moves, gets their opponent on the ground, they sometimes stomp on their opponent's head. Their adversary is absolutely not a threat anymore, yet the winner will continue to bash on the head of the unconscious person. This is exactly what happens in the chimpanzee community. Even, in sometimes random, unprovoked circumstances, a person will commit an act of aggression. The destructive aggression does not have to be physical harm, some people really enjoy, they really feel good emotionally, when they sarcastically ridicule another person with words. Still, this "enjoyable" behavior to the individual perpetrator, is viewed as "bad" and destructive by most rational people. Sarcastic, verbal ridicule, unwarranted aggressive verbal abuse, probably happens more often than physical abuse. Both words and behaviors can be constructive or destructive. The aggressor thinks they are doing "right" in taking revenge, getting even, getting justice, or just to be entertained by their own aggression. These are moral decisions, even though they are destructive moral decisions. Whatever adrenaline filled choice is running through the destructive person's mind, at that moment, they are thinking that acting destructively is a right and a "good" thing to do. Destructive actions "feel good," this is apparent, since the perpetrator is choosing them. The fight winner is violent, and "stepping" on another person's rights to a safe and comfortable existence. The aggressor is the destructor. The aggressor is getting some benefit from these dreadful acts, whether it is physical, mental or verbal destruction.

When I was a principal in an elementary school, snow would fall. The students would get extremely excited and go play in the snow during recess. There were two morning recess sessions, the first session would go out and build snowmen, and snow forts. Some students would have spray bottles of dyed water and paint images in the snow. Every single time, after the "creator" kids were done, other kids, at the second recess, would come and destroy what had been made. The students coming out for second recess would *always* knock apart the previous recess' building activity. It is natural, to get emotional enjoyment out of both creating and destroying.

In the ninth and tenth centuries, Viking raiders crossed the North Sea with the intent to slaughter and take land. There was no thought other than the Norsemen goal to kill existing residents, and take slaves, and other items which were not their own. Vikings/Norse people wiped out entire areas of modern-day Ireland and England. Similarly, when the United States government finished its Civil War, generals and troops were sent west to subdue, and annihilate, the Native Americans living in the western United States. In this case, power, destruction, and killing for selfish gain, was the sole intent of the U.S. government. Different groups and cultures, now, and throughout all history, place a high value on aggression and killing power. The Spartans of ancient Greece, Viking culture, European colonizers wiping out aboriginal people of Australia, the U.S. obliterating the Native Americans, Romans, Nazis eliminating Jews, Greeks, Mongols, Celts, English, Persians, Saxons, Macedonians, Saracens, Normans, Native American tribes (versus each other), Assyrians, the Zulu Tribe, the Japanese destroying the Chinese, "Christian" Serbians who systematically murdered and raped the Muslim and Croat population… These are just a very few instances of the many destructive/aggressive actions in historical narrative. Let's get real here, this list could go on forever. Knowing and utilizing how to kill and/or dominate others happens all over the world, throughout all of history, groups/nations have had this lion versus gazelle mentality.

Killing, and aggressive imperialism is everywhere, all the time, but is it right? Is an effective killing soldier in war a hero, or not? It all depends on your own point of view. Rome (Eastern and Western) imposed her ideals (and killed plenty) for over 2,000 years. In 313 AD Constantine, the mightiest warrior, with the mightiest force, decided Christianity was not only now legal, but also beneficial. Not only was the murder of Christians not completed anymore, as it had been by previous Roman emperors, but by 325 AD, emperor Constantine was calling for Councils (Nicea) to get Christian teachings straight. In a matter of just a few years, Christianity went from a maligned side-note, to the Roman state religion. Ironically, "might makes right," was needed to install the pacifist teachings of Jesus. Dwell on that, for just one oxymoronic second: destructive, aggressive and deadly power, was needed to install Jesus' pacifist teachings. The fiercer, and/or the technologically superior group, rules the inferior group. If there is a creative entity of this world, it enjoys and values aggression and hostility. It is built in, and readily perceivable.

For better or for worse, the USA is Earth's current "Rome." In both a physical sense (power) and an academic and technological sense, the United States, currently, is probably able to impose her will if she wants. Is it right or wrong? Ask the Native Americans, you'll get a different answer than the settlers. Is it right that a sniper for the USA, Chris Kyle, killed many Iraqis during the U.S. invasion in the early 21st century (Dale, 2012)? Ask Iraqis and Americans, you'll get a different set of answers. Should the United States, or other global powers feel guilty about being powerful? The variable views on Christopher Columbus are another great example of the different views on aggressive imperialism. If a Norseman died with a sword in his hand, fighting an enemy, he could enter Valhalla - the heavenly court of partying and fun. A Muslim can earn the title of "Shaheed," an honor for those who die killing infidels. These jihadists believe they will go to the highest level of heaven for killing people that don't agree with what they believe to be right or wrong (Kamber, 2001; Victor, 2003). Many people residing in powerful countries do feel guilt is warranted. Others are just happy to be in the position of power. Objectively, having power is not right or wrong, it is the perceptible reality of how our realm works. What is deemed as right or wrong is clearly up to each individual's viewpoint.

This imperialist approach continues to this day. The reasoning could include these following ideas:

-You are not similar to me

-I do not know you

-I don't agree with / I dislike you

-I want what you have

-I'm stronger than you either physically or technologically

-Might makes right

-I will annihilate you

These are all morals by which many people live, still, to this very day. These are not the virtuous morals that build, these are the beliefs of the survival-of-the-fittest standards of behavior. Might makes right, and the rule of the technologically advanced. In short, this is an imperialist mode of operation. Here is a hard, but honest reality of our world, strength and superior brain/technological power will determine who rules. The people who are stronger in both an academic and/or a physical sense, are the people that have the power to impose their beliefs. Is it "right?" This destructive imposition of whatever other moral paradigms people may hold is a fact of our existence. Viewpoints are variable, "wicked people" that are annihilated, (or the annihilator, depending on your viewpoint) are only wiped out because they are technologically and/or tactically inferior, *not* because they are detested by some God or gods.

People get very upset, they "know it is wrong," when acts of aggression are perpetrated. Non-empathetic people that crush the rights of another person are viewed as evil. Especially when a child, or any innocent person, that is not able to stand up for themselves is being oppressed. Why do people feel the same indignation when they see the unjust degradation of another person? The people who are actually 'mowing over' the innocent… do they, or do they not, feel the pull of their conscience? The probable explanation is that they don't prioritize values in the same way, especially during each, specific incident. The aggressor rationalizes atrocities, or the taking of the rights of another individual, because they don't prioritize their morals the same way as others. The brutal/aggression moral is more valuable than the compassion/helping moral to some people.

Aggression and hostility happen right alongside love and compassion. The alligator will drown a human to tenderize the meat and eventually eat. This is the same alligator that will hold its mouth open to allow feeding birds to pick its teeth clean. In complete juxtaposition to the destructive mode of operation, there are people who are doctoring, helping, and educating others. These are **constructive** actions completed by people who are builders. Some people give major amounts of time and materials with the sole intent of helping others. There are individuals who spend their lives here on earth building up others' quality of life. These two approaches: constructive and destructive, just as we see in the scientific study of our natural world, are in a continual dance, all the while going toward a stable equilibrium. If the destructive mode of operation version of ourselves, and our society, were to be continually utilized, we would cease to exist (Aburto & Beltrán-Sánchez, 2019). To continue existing there is a perceptible balance, a sort of pendulum effect, a homeostasis of actions. This is exactly what we see occurring in the natural world. The destructive and constructive variables in our lives act in such a way so that the overall conditions of existence remain fairly constant. The oppositional balance between construction and destruction are needed to define each other, and to balance each other.

So far, we have predominantly focused on the destructive side of things, now, let's look at the counterbalance: constructive behavior. There is a video where a long-necked swan (geese do this as well) would take food pellets from a trough in its mouth. The swan would turn with the food in its mouth from the trough, open its beak, and feed fish that were swimming close by. The food was put there by generous humans for the swan. Unselfishly, benevolently, humans give food to animals. Then, unselfishly, benevolently, the swan would not take the food for himself. Instead, the swan, in turn, gave it to the fish. The swan would turn and feed the fish who were happily benefiting from the swan's charity. There are moments, even in nature, when at a cost to an animal, the animal will help a fellow creature (Riolo et al., 2001). In the same way that destruction naturally occurs, goodness/kindness, and building/helping does occur naturally in this earthly realm as well.

Screenshot credit: jammingontheone1. (2018) YouTube.
https://www.youtube.com/watch?v=I3tosN66sLo

There are people that will travel to areas where others are poor and in need of medical and/or nutritional health. These kind-hearted people will help cultivate land, and dig wells, for those in need of food and water. They will build houses for the poor. There are compassionate people who will devote time with the sole intent of helping others. It feels good. It emotionally feels good to give presents to others. It feels good to impart knowledge in order to make another person's quality of life increase. This is why education and having general knowledge is so important. It feels good to build others' self-esteem with genuine compliments. It is emotionally rewarding to help other beings with basic needs. These are just a few of the ways "a builder," a predominantly constructive person, lives their lives.

What holds true in the world of scientific study (as discussed earlier in this book), homeostasis, leveling, that same phenomenon holds true for constructive and destructive behavior. Mandelbrot Set, animal populations, and atoms losing neutrons and protons in an attempt to become stable. Constructive and destructive occurrences "level off" in the same way as in the realm of emotions, in the same way as in the realm of science. These few examples could be applied to thousands of situations. The destructive/selfish element will always exist along with the constructive/philanthropic. Homeostasis and balance happen over time. "Destructive" only exists *with* "constructive," the pendulum swings, but there is always an equilibrium position.

Constructive behavior and destructive behavior, whether it be nations or groups of people, or individuals, these two aspects are always present. The very same people that were builders a minute ago, may become instruments of destruction. Who is continually philanthropic? Who is continually destructive? What nation, or person, is continually philanthropic or destructive? None. When does the builder become a destructor? And vice versa? Destructive and constructive behavior is completely determined by each individual's opinion. A person who is making charitable decisions can, and often does, turn into a destroyer if their way of life is infringed upon, or some other perception of unjust circumstance arises. A builder, an unselfish, kind person, can, (and does at times) change into a vengeful destroyer. The same holds true for individuals, groups of people, and nations. Actions on the destructive/constructive continuum constantly change, and they are viewed differently, by different people.

Societal priorities change over time. Currently, being destructive, or aggressive is considered to be wrong where I live in the USA. In my current society, with emphasis on the immediate squashing of any hint of bullying, is practically the opposite approach to a Viking, a Jihadist, or a Spartan. This was a common aspect of my job as a public-school principal, to stop aggression between children. I was to stop and resolve conflict between parents and teachers, teachers and teachers, students and students, and all combinations of these participants' disagreements. When I worked as a school principal, aggression was definitely viewed as being wrong. However, this non-bullying emphasis is not universal throughout my current culture. Here in the United States, there are certain settings where aggressive, imperialistic behavior is wanted. If you are playing American football, it is good, wonderful, and even expected, to be aggressive, and to dominate your opponent, to "destroy them."

Chapter 21

Ubiquitous Continuum #3: Freedom/Restriction Basic Rights to Your Own Body

As has been covered earlier in this book, people seem to have freedom to make choices. We can see others making choices, we have the ability to choose what actions we consider to be beneficial. In, in order for any person to be culpable for any decision, every person must have freedom of choice. Let's look at what is considered by many to be a basic right… freedom. Let's say each person should have the freedom to choose what happens to their own body. Where your body travels to, to make your own life choices, the basic freedom to do with your own body what you want. Can we say that this very basic, personal right, should and would be a good idea by which everyone should live? Almost every sane person would say, "Yes!" Yet, even this most basic right would be considered unacceptable by many people. For example, take the Governor Jerry Brown's mandate that all Californian children be immunized (Nyathi et al., 2019; Martinez & Watts, 2015). This California governor banned all personal, medical, and religious exemptions to force parents to give vaccination shots to their children.

Pondering Our Existence for Practical Living

There are many such incidents where people do not have choice as to what to do with their own body. During the Covid-19 pandemic, everyone's movement was restricted, especially when vaccination proof was mandated for travel. People were fired from their jobs if they did not get the Covid-19 vaccination (Hsu, 2022). Laws as big as deciding whether a woman can get an abortion, to something as simple as mandating people who ride in cars to wear a seatbelt, *these are all infringing on each person's decision for their own physical body*. You do have a say in whether you wear a seatbelt, or not, in the U.S., but you will also get a punishment. A ticket will cost you money if you choose not to wear a seatbelt. You could choose not to get an immunization, but there were severe consequences. There are all kinds of restrictions on people who smoke. There are laws made deciding for us what food we are allowed to eat. These are all examples of hindrances to having rights to one's own body. Since everyone has a different idea of what is right/good/beneficial, as opposed to what is wrong/bad/harmful, logically, these restrictions on our own personal body should not exist if freedom to one's own body is valued. The higher value, in many peoples' minds, would be to have the freedom to choose what happens to their own body.

Historically, women have not had the same rights as men when it comes to what happens to their own body. Up until very recently, women had to stay in the home. Historically, women in the United States were not allowed to work, or go to college (Cox, 2024). Who is responsible for women not having the same rights as men? Does this mode of operation stem from embedded, ancient, gender roles? Is it merely a concocted, cultural pressure to limit the rights of women? In certain Muslim societies today, women must obey, completely, their husband's will and wishes. Even to the extent that women's sex organs that give pleasure, are removed (Isa et al., 1999). Women must wear the clothing prescribed to them, they must cover up, even their very own faces. There are many societies that enforce this women-must-cover-up-rule known as 'purdah' (Korotayev et al., 2015). Purdah in strict Hindu and Muslim areas of India, and also the Taliban in Afghanistan, are just a few instances where women do not have the basic say over their own body (Shaheed, 1986). Will there always be some invented reason to take what seems to be very basic rights away from a person? It is easy to see some "God ordained revelation," or blame traditional roles as cultural dominating pressure, but this is not beneficial to certain segments of society, in this particular case, women.

As we can see from the examples above, currently, people having all the power over their own bodies is still not "allowed" by governmental rules and cultural pressure. Slavery is an overt example of living with limited personal decisions over one's own body. Slaves have very little freedom to decide what happens to their own physical body. Slavery continues to this day in many parts of the world (Howell, 2014). Slavery, in the past and present, is in complete opposition to my personal role as a protector of people's rights. Almost all, rational, modern people feel slavery is repulsive and wrong. Contrariwise, slave owners think it's right, good, and beneficial, to own someone else's body. To use the other person's body however they want. Pimps and whores have this same relationship; the body of the whore is at the whim of the pimp. This setup does not coincide with most people's value system, but it is the chosen value system of many people. There are slave owners to this day in Eastern Europe, SE Asia, and Africa that still value slavery as a benefit (Howell, 2014).

In Nazi Germany during the 1930's and 1940's, if you were of Jewish ethnicity, you were murdered. The Nazis could not kill the Jewish people fast enough. Either death, or slavery in work camps was where Israelites and other dissenters' physical bodies were forced. During World War II, in the United States, people of Japanese descent were forcibly sent to internment camps. Men in the United States from 1940

– 1973 were drafted, forced by conscription, to go fight in wars. Even during the Civil War in the U.S., men were forced into military action. It is ironic that the people fighting for a person to have rights to one's own body (freeing slaves), were fought for by people who weren't given the rights to their own body (forced to be a soldier by conscription). Union soldiers, many conscripted, fought to free the slaves in the southern United States. 360,222 northern soldiers *lost their very lives* fighting to give southern slaves freedom (Zeller, 2023). These are just a very few historical incidents to show that throughout history, up to the present moment, many people have not had rights over their own body.

There are many slave owners to this day who somehow justify their slave owning practices (Global Slavery Index, 2023; Hamper, 2024). The Bible has been used to legitimize slavery in the United States (O'Brien, 2023). Will the slave owning Romans of antiquity be punished in the afterlife for owning slaves? There were countless slave owning groups throughout all of history: the Norse/Vikings, Portuguese, Spartans, Nazis, Saracens, Native Americans, just to name a few. These people did not think slavery was wrong enough not to partake in the practice. Most people today, around the world, agree that owning another human is disgusting, it's wrong. However, our judgment stems from our upbringing and culture, right and wrong, what is sinful, and what is virtuous, (as seen throughout history) is clearly in the eye of the beholder. Astonishingly, this includes people having the very basic right to their own body being withheld. Rules and laws come from groups of people (government, religious, or other groups) who set a certain action as a priority that must be followed. When these laws/rules are made, invariably, someone who has a different priority, a different idea of what is right and wrong, will have their beliefs and actions stopped, or placed in an inferior position. Any act in opposition to local institutionalized behavior and beliefs, may even be considered "wrong," or "illegal," by the government or local society. Sadly, it was illegal, to help slaves escape from the southern United States during the 18th and 19th century (Rickert-Hall, 2020). Every person is invested and concerned about right and wrong; this is a universal human trait. All humans have opinions, emotional contentment, or contempt, when confronted with any, and every situation. Whatever situation we have to deal with in the moment, your personal view may, or may not, coincide with beliefs about even this most basic right of how much authority people have over their own physical bodies.

Can we even imagine what society would look like if we all followed this one, very basic rule - Every person has the right to do what they want with their own body? No more rape, or murder would occur. No more striking another person. No more slavery, conscription for war, or any act that takes away the rights that infringe on the power one has over their own physical body. Frustratingly, this scenario is impossible in practical application. For the same reason we can't have a society that only performs kind actions, the same obstacles apply to this basic right of self-body authority. There will always be somebody else's "moral," or "correct," teaching that stop rights to one's own body. Laws about clothing, laws about what food/drink is best for the human body, (medical and safety) these will continue to hinder this seemingly, most basic right to occur.

Chapter 22

Hierarchy of Virtues
Intellect, Confidence, Self-Control
and Physical Prowess

Is there a better way to live this life? How do we approach everyday decisions to achieve success in some perceived hierarchy of what is valuable? Pondering what is right and wrong, what is constructive and destructive, has been our focus, now, let's examine practical living. Are there actions we can do to have a flourishing life? Does a great life mean pleasing a God or Gods? Does a great life mean devotion to some state government? Does a great life mean having the freedom to do what you choose, when you want to do it? These are just a few ideas that people consider to determine the ingredients of what makes a wonderful life, and there are plenty of ideas and differences.

What is interesting, is that 2,400 years ago, ancient Greek philosopher, Aristotle, identified many of the same attributes and qualities as being good, or desirable, as we would today. Acts concerning courage, happiness, being victorious, gaining wisdom, loyal friendship, self-restraint, these same attributes seem to be universal since they are same over millennia. Going way back to 750 BC, Homer wrote the Iliad.

Is it not incredible that when we read about Agamemnon's greed, and his bullying tactics, that we automatically know they are bad characteristics… almost 3,000 years later? Staying with ancient poetry, in the Iliad, is it not uncanny that we admire Hector and/or Achilles' courage and toughness? Paris' cowardice is derided; Helen's beauty is an impressive source of vast power. Here is a short list of the right qualities of thinking/acting/speaking correctly, as outlined in eastern religions: selfless service, truthfulness, not stealing, honesty, knowledge acquisition, meditation… which of these would not be important in western religions as well? Considering thousands of years of separation and cultures from every part of the globe, it seems that the same virtues are predominantly considered praiseworthy, or shameful, everywhere.

Picking a best way to live life may be determined by deciding which personality attributes are most desirable. "Good" or "bad" qualities are what we humans pick to admire and aspire to be, or to deride, and try to avoid. What are good and bad attributes? Are people who do not have the personal attribute of self-control wrong/bad or right/good? How about courage, beauty, kindness, cunning, selfishness, laziness? Listing some of these personality qualities, most people would have a reaction, a judgment, of which attributes are good or bad. People who live by what are considered undesirable traits may be labeled as, "losers," "sinners," and/or "uncontrolled," etc. It is interesting, that all people, will at times, choose attributes that are usually frowned upon. Some of these, generally undesirable traits include: selfishness, laziness, trickery, violence, persuasiveness, manipulation, cunning, and aggression. Possibly, we pick to live by these 'less admired' traits on certain occasions because of the life situations we encounter, or, it's just beneficial in a certain situation. Also, some people may have problems reasoning, some may have hormone disorders and/or a traumatic childhood that encourages them to pick attributes that are more in line with selfish gain. There are many reasons we humans decide to act in ways that would often be judged as utilizing "undesirable qualities."

I was teaching sixth grade science in the 1990's. During one of my classes a student was relaying an idea from a show on television. She said talking to plants made the plants grow better. The class discussion went to whether speaking kindly, or rudely, to a plant would make a difference in its growth. Would the plant that was constantly berated and ridiculed wilt, or not grow as much as a plant that was given compliments? We decided to control variables and get two plants, the same species, about the same size, and place them the same distance from the window. We watered the plants the same amount as well. One plant was named Doofus; the other plant was labeled Cutie. Only ridicule, and rude, cutting statements were to be made to Doofus. Cutie could only have compliments and kind things said. The one (unforeseen) variable we could not control was the overwhelming power negativity has over the human heart. I had thirty-one students in my class, when the students spoke to the plants, there was a line of students waiting to yell rude things at Doofus. On the other hand, hardly anyone wanted to give kind compliments to Cutie. Even after I discussed with our class that we had to give *equal* amounts of derogatory statements and complimentary statements to the respective plants, the students could not do it! They would yell four or five ridiculing statements to Doofus, "You ugly! Piece of crap! Disgusting plant! You suck! I hate you!" Then, they would quickly go to Cutie and muster a single compliment such as, "You're pretty." Even after I told them to keep the comments equal between Doofus and Cutie, they still had many derogatory things to say to Doofus, and not much at all to say to Cutie. The same phenomenon is easily seen on any social media platform, it's easy for us humans to be negative. It is difficult to suppress this quality.

In this section of examining whether there is a hierarchy of virtues, our ability to use logical intelligence to reach possible conclusions about our lives, this is a paramount virtue. Reasoning is healthy, beneficial and viewed as a "good" virtue by nearly everyone. The use of our academic mind is the basis for all advances in all other aspects of life. So, this book's goal is virtuous, because we are using reason, and science to decide the big questions like, "Why am I here?" "Is there a purpose for this life?" "What is my Purpose?" "Is there a God or Gods?" "Is there an afterlife, a place where something of ourselves carries on?" If gaining wisdom is a worthy virtue by which to live, what makes a person wise? An intelligent person: understands, makes predictions/wise choices by analyzing situations, connects concepts, truly listens, processes and remembers information, and understands varying viewpoints.

In the case of morals, whether they be viewed as healthy, or detrimental, it seems we pick what we think is best at any given moment. We all encounter internal urgings. There is definitely some urge felt in our hearts, to act, to think, and to choose this way or that. We listen to those urgings; *we should* listen to those urgings. Sometimes the moral to follow is in the vein of selfish gain, sometimes the moral to follow is a virtuous "constructive" moral. To be specific, some person's chosen priority of personality attributes may be loyalty. It feels good to this person to be loyal. They perform actions, normally, that demonstrate loyalty. One could pick any mode of operation: building others' up, equality, perseverance, etc. These actions that align with a person's prioritized qualities, feel emotionally good.

When we pick certain virtues to live by, does that make a person superior to the person who has chosen different virtues? Some would say they are superior because they live by this, or that, preferred virtue. One thing that is true, you are defined by others by what you repeatedly say and do. Your chosen actions and stated beliefs, (that are guided by your inner chosen qualities) define you as a person. If you want to be a certain way, make a promise to yourself, and keep that promise by choosing actions, thoughts, and words that support your chosen virtues. Your own view of yourself, in addition to the views of others who have knowledge of your existence, is created by your chosen attributes and actions. That is why, when one person describes another person, they state what work they do, and what recreation activities in which they partake. Each person is often generalized in categories such as the redneck, the smart guy, the rock star, the nerd, the druggie, the whatever… Did we take on these roles at the suggestion of a commercial? Did a friend influence us? Our parents? Advertisements work; that's why truckloads of money are spent convincing us of what is valuable. Commercials urging us to define ourselves, most often through things they can sell us. Of course, our chosen priorities will always be influenced by what we have taken in through our life experiences. A superior way to live our lives would be to follow a rational, analyzed, intrinsically created approach to life. Our personality traits reflect chosen and prioritized virtues. "She is an honest person," "He is a confident guy," "He is conniving," "selfish," "fun," "bold," "talks to much," "long hair," "has tattoos," "is timid," "hard-working," "looks short," "skinny," "walks fast," "enjoys science," "gets frustrated easily," ad nauseam. The important thing is, yourself, and people around you, see your actions as choices in prioritizing which virtues are important. Then, those actions and virtues, along with your physical attributes, are the ingredients of you. A eulogy, or an obituary, are exactly this, they are the description of what you picked as an important way to live your life.

What people see you doing, saying, and deciding, gives the observer descriptors to define you. You, know your own inner motives, so, you may not think of yourself the same way as other people think of you. It's very difficult to throw off what other people say and think about who you are, these prevalent descriptors, that we may hear over and over, do tend steer our actions. The true reality is, the labels of others, may be true, or they may be false. Whether the descriptors others us to describe us are true, or not, it's very difficult to be different than these labels. A school student who has broken the rules in the past knows he is "the bad kid." The kid who gets his essay back with red marks all over it, knows he is not good at writing. "I am bad at math," is such a common statement. This is why, because someone had to correct their wrong math problems when they first learned a new concept, and the student learned to take on this descriptor of themselves as being inept at math. Self-fulfilling prophecy is a powerful force (Chandrasegaran, 2018). The labels of others' may not be true. Possibly, developmentally, the person was not ready for a certain math problem. Maybe, the person didn't put effort into solving the math problem. They might be the greatest mathematician of the age, but someone corrected them at some point, and it's hard to shake off other peoples' labels. We do what is expected of us the majority of the time. Pressure of society, especially individuals living in close proximity to us, are a major source of our defining characteristics, and how we think of ourselves.

Sometimes circumstances have influence on our individual happiness. For example, a tyrant, or a tormenter can come along and ruin our lives. However, if a person was never cognizant of all the possibilities to have a fulfilled life which *could* include: travel, freedom, less pain, more pleasure, material possessions, and physical ability, then it is possible the person who is limited by said tyrant, could still have the feeling that they lived a fulfilled life by following their chosen virtues. If others judge in comparison to other people's lives, there will most likely be a perceived area of deficiency. However, if they only know their own life, not knowledgeable of their limitations, this hypothetical person could be very happy. A proper comparison could be space travel. For us, living in 2025 on planet Earth, we don't know the possible, amazing, and incredible joy to be had by traveling to another planet. Since this space experience is not even an option, there is no angst, longing, or sadness for not having this capability. This is no different from the fish trapped in a tank, or a peasant trapped on the land of his baron. Happiness can be found in a life even under strict limitations. Each person judges, from their own viewpoint, which life is a flourishing life, and which is not, there is never an unbiased perfect judge. Everyone is just a person with an opinion. A quiet life by the ocean, staying with chosen personality qualities, subsisting by catching fish, living simply in a shack, is not inferior, or superior, to a billionaire vacationing around the world. What is considered flourishing is idiosyncratic.

Our chosen virtues are interwoven with our personality traits; these can be viewed as either beneficial or detrimental. A single attribute can be viewed in different ways. Let us look at confidence. Many ancient philosophers speak about courage and confidence, which are closely linked. Is it good for a person to be confident? If a person feels they will be successful in whatever endeavor, this helps in the positive attitude along the way. A person may have courage to do something because they are confident in that area. Most people would say having confidence is a good virtue, but if the word "arrogant" is substituted, most people would say arrogance is negative. What is the difference between arrogance and confidence? Arrogance has the element of direct competition, an arrogant person brags about themselves, trying to prove to others they are important. If other people brag about a person, it is fine and celebrated. If a person brags about themselves, they are arrogant and maligned. This single virtue of confidence is thought of in differing ways.

In practical living, it seems to ring true that every undertaking is best completed with the virtue of moderation. Moderation is a generally agreed upon, 'good' virtue by which to live. Hence, all the previous points in this book on equilibrium and homeostasis. Certain drugs and alcohol taken in moderation brings joy, tranquility, and lightness of heart. At the same time, alcohol and drugs in excess will physically, and mentally, destroy a person. Addiction may lead the overindulgent person to perform *very* destructive actions. The same goes for food, overindulge, you will be obese. Inordinate amounts of food are unhealthy, and will lead to an early death. Conversely, food in moderation is necessary, and brings joy and health. Could the same be said for courage? Everyone would say courage is good, but too much courage will lead to stupid and reckless actions. The same relationship can be made for something as basic as the virtue of kindness. Kindness is good, but excessive kindness can lead to the destruction of one's self. An overly kind person would always be at the call of others, putting up with others' disrespect, or even abuse, all in the name of being kind. Kindness in excess can be destructive. Even these societally accepted, "valuable," virtues need to be completed in moderation.

It may also be beneficial to have moderation in emotion. How we feel our own emotions, as well as showing what we feel internally to others, yes, moderation applies here too. There are people on either end of the spectrum, excitable, loud, crazy, crying, laughing… big emotional reactions at the slightest incident. On the other end of the spectrum, there are dull people, no smile, no emotions. It seems that emotional self-control is healthy. More impactful than what we show, outwardly to others, is how we direct our emotions internally. It is difficult at times, but it can be done. People on the showing-emotion side of the spectrum, often, have to tell themselves, "Stay calm, don't react, don't answer right now." As in all virtues and personality attributes, moderating and controlling your inner emotions, and outward display, is just as valuable as moderating other areas of life. It can be debated as to whether outwardly staying calm, having self-control, is always beneficial. There are examples of when living by unstrained, reckless abandon, has been impactful. There are many individuals throughout history, but these few immediately come to mind: Catherine the Great,

Pondering Our Existence for Practical Living

Genghis Khan, Janis Joplin, Alexander the Great, Queen Elizabeth I, Napoleon. These historical figures had productive, certainly impactful lives. Yet, they had little to no self-control whatsoever. Even on a physical level, others come to mind for living a fulfilled life, but living with very little restraint. My son, Stephen, and many other skiers, drop off cliffs, or surf hundred-foot waves that could crush them. They perform incredible physical feats, the ability to allow one's body to flip/somersault/contort, shows complete unrestraint. So, self-restraint is not *always* the best approach, but it is generally helpful to live with moderation.

As we see with looking at self-control, acknowledging a hierarchy of virtues presents many problems – it's muddled. Self-control and moderation are closely related. Who determines which person has self-control? What is the right amount of self-control? In what activity of life? Earlier in this book, when discussing changing ideas about what is right and wrong, I wrote about uncontrolled impulses that are currently acceptable. A religious person, who is completely addicted to reading their scriptures, could they be accused of not having self-control? How about an addiction (no self-control) to church activities? Or an addiction to looking beautiful? Or an addiction to some academic endeavor? When do you ever hear someone say as a criticism, "You have no self-control when it comes to reading and gaining knowledge!" What amount of self-control is the right amount for activities some would say you could never do enough? How about the inability to control one's own judgment of other people? These are addictions where some people may just have no self-control, yet society chooses what is a "good," or "bad," addiction and/or personality attribute. A person will still feel happier, and more fulfilled, completing what **they** decide are valued actions and personality attributes. Even if there is outside pressure to do something other than what your inner emotions are guiding you to do, one should still follow their own inner steering. This is how other peoples' views, and society's views, change. Plus, a person will never be fulfilled if they are merely following someone else's idea of what is a good personality attribute, or a worthy life pursuit. Stay with what your own senses tell you, then you will be true to oneself, and to those with whom you interact.

Any chosen virtue, including self-control and moderation, relies upon the fact of a person having a choice of actions. You are the one with the ability to direct your own behavior. As to how much your behavior is affected by other people is idiosyncratic. This demonstrates the very important idea that all virtues, personality traits, physical traits, and actions, are viewed differently by different people, especially under the label of being right, or wrong/beneficial, or detrimental. When a person is dogmatic about their prioritized attribute, they sometimes attempt to assert their attribute as being superior, or the only *important* attribute. The reality is, no person's descriptor is truly "superior" to any other. These preferences are arbitrarily foisted on us, and accepted by us, through local culture/family, and/or stories/music/entertainment and advertisements. Is there a way to define ourselves and others without personality descriptors? It would be impossible, however, the groups we become enthusiastically identified with as the best, truest, "only way," to live, can have a severely limiting impact on the many possibilities of our lives. Many people will "outwardly believe" what they know to be untrue internally. This is done to fit in and decrease one's anxiety. It's easier to give into the pressure of your local community, than to live with different priorities, or be involved with what is seen as undesirable groups.

Usually, the way to what would normally be considered a great life, is based on acting a certain way. Certain, chosen attributes, or qualities, are definitely part of an amazing human life. One of my colleagues did an activity where we thought of people we admired, then listed the qualities of these individuals. We picked people from history, sports, spiritual realm, family, and characters from fictional books/movies, that we admired. The first interesting discovery is; the qualities we admire in these characters/athletes/leaders, turn out to be a perfect description of ourselves. The psychology has to do with the fact that the qualities we admire, what we strive to be... we are. More importantly, when people (and I have asked lots of people to do this exercise) certain human qualities seem to be popular and constantly admirable. Hope and resiliency, these are elevated as incredible attributes, and needed for survival. Over and over again, attributes that comprise a great person include: Perseverance, kindness, determination, toughness, honesty, humor, generosity, being skillful, intelligence, being selfless, supportive, humble, skillful, artistic, determined, innovative, passionate, helpful, good looking, and hard working. Why don't people admire the opposites of these popular qualities? Virtually no one said lying is an admirable quality. Overall, people are consistently, naturally, bent toward egotistical selfishness, but yet, it feels good to be (and is celebrated) generous, and to help others. Why does it feel, emotionally pleasing to help others? Why do people say they strive for, and prefer generosity over selfishness? When do you hear people brag, "I'm a truly selfish person!"? Hard work is lauded and is often the reality behind "a miracle." Who boasts, "I'm a lazy person!"? No one said, looking ugly, being weak, boring, sad, apathetic, harmful, divisive, selfish, or lazy, were desirable attributes. Why not? This goes back to the section previously discussed, virtues are the admired descriptors that help society to function. The descriptors that are not desirable are usually used for selfish advancement. It is the same reason as why stealing is detrimental. There are certain personality qualities that are viewed as good, longed for, and praised. Conversely, there are other personality qualities that are predominantly viewed as wrong, bad, or to be avoided. The morals often viewed negatively by which we, at times, choose to live, are valuable for personal gain. Certain attributes seem to be more beneficial, and positively celebrated because they help societal living.

People tend to find those who have similar choice in personality attributes. Some people place being dominant (sometimes physically) over another person as an important value. This person will act in such a way to further this clique group identification. The person who decides being physically tough is important, might bully other kids at school. They might get in a boxing ring, or a mixed martial arts combat group, or play other combative sports like rugby or American Football. In social situations they might physically shove another person. In contrast, another person decides the priority of being intelligent as an important value. They will read a lot, live in front of a computer, take notes, focus on research and study in some specific area. In conversations they will try to impart their knowledge to others, join book clubs, and other academic groups. They may verbally debate their knowledgeable beliefs with others. The choices of actions, and who this person interacts with, will be vastly different than the person who values physical toughness. Groups are made because people find others that have common values. It seems most people pick labels in their tribalistic efforts, "I'm a Democrat," "I'm a Christian," to give themselves the all-important comfort of belonging to a group. People define each other, and themselves, by labels of groups. These groups are formed by people with common value prioritization. It is valuable to see the limiting factor on our existence when we limit ourselves to a group label. Not seeing your own value preferences as possible "blinders" can limit our quality of life. Can a person not be tough and intelligent at the same time? Simultaneously artistic, ambitious, and outgoing? Too often, we live with tunnel vision, convincing ourselves, and others, that we are playing a specific role. Sometimes we are corned into a role by the comments and actions of other people.

Pondering Our Existence for Practical Living

What are the good/pleasurable things in life? Wine? Study? Scenery? Conversation? Love making? Fast cars? If pleasure leads to happiness, and happiness is a good and worthy pursuit, is there a hierarchy in what brings pleasure? Is there is a hierarchy in which virtues (or actions lined up with these chosen virtues) are beneficial or detrimental? What makes you happy on this very day? Is wanting a fancy car not as good a pleasure as having a deep conversation? Is a great tasting meal as important as creating a sculpture? Each individual will have a different list with varying priorities. The important thing is, none of these hierarchy lists are wrong. Whatever actions bring you joy... do that!

As we progress through this life, our views and perceptions change. The things we enjoy, dislike, are embarrassed about, are excited about, our sense of right and wrong, and what we think has value, all those thoughts change. What is important is greatly affected by the teachings of our surrounding family and culture. Most people do change over time, as they have more and more varied life experiences, and interactions. I am also sure there are some people that are so indoctrinated, and they have invested major amounts of time and money, into a particular world view, that they do not want to change. Often this is the case, where it is too embarrassing for some people to admit believing in whatever fictional religion, or belief, in which they have chosen to believe. They and their family, for who knows how many generations, have invested time, and money and 'their very souls' to someone else's beliefs. It is true that in some cases, even people with passionate, steadfast, entrenched beliefs, have the ability to change if they learn about, or experience something, that changes how they see the world. However, it is very difficult. We take pleasure in doing what *feels* right to us. We need to do what we each feel we are meant to do – and do it well. However, performing actions of desired virtues are no guarantee of happiness, it's just the best to-us-in-the-moment route. Living by our chosen virtues make us the best humans we can be.

Living for joyful/pleasurable moments, to do things that make a person cheerful, may not make the character-trait-list of courage, or moderation, but it is equally a logical quality by which to live. Many people reading this may dismiss the idea that seeking pleasure in this life, and/or living a fulfilled life (the best they can) is not important at all. Many religious people may put the primary goal of this life as doing actions that follow others' presented beliefs, to put all their effort in what may come *after* we die. It is possible, the unseen, unprovable, hoped-for-spirit- world could be viewed as vastly more important than seeking pleasure in this life. It is possible that the many, amazing iterations of paradisiacal afterlife scenarios could be measured as much more important than any chosen individual life purpose. The purpose for the person living for a rewarded afterlife, becomes pleasing the god, or gods, that place people in the desirable afterlife scenarios. The reality is, this approach includes the drawback of accepting the whims of other people's stories and ideas. Living for pleasurable experiences pays off now. These religious stories may, or may not be true, religious stories are a remote possibility, at best, and only ever just a chosen belief. It may be more valuable to live for happy moments in this current life.

Chapter 23

Interpersonal Relations, Communication & Friendship

Another virtue would be to be amicable, to have the ability of creating skillful interactions with others. How one accepts, or not, encourages, or not, and acts towards other people, has a massive impact. Considering interpersonal interactions and friendships, is it valuable to even have friends? Is maintaining a friend, and having interpersonal relationships, a beneficial and worthy life purpose? A few trustworthy friends *may* ease suffering in this life, and people may feel true joy when they interact with a friend, but there are drawbacks to friendship. There is an element of freedom in being a loner. Learning to value being by one's self, and truly enjoying times of solitude, can be as rewarding as friendship interactions. There is the fact that no one is forcing you to do things *they* want to do. Not having friends frees you from having to listen to others' thoughts, or praise, or criticisms, that may cause you stress. The placement of how important friendships are will most assuredly vary from person to person, just as whether to be friends with people of differing beliefs will also vary. The reality is, it is possible to get along through life without friends. When I counsel with young people, thousands of people over the years, friendship is a common problem for many. People have real difficulties with who is expected to be a friend. To help people figure out the basis of who is truly a friend, it's a clarifying activity to define the actions of both friends and enemies.

To define what actually constitutes someone being a friend, it's illustrative to define the opposite: what makes a person an enemy? An enemy says nasty things about you, they spread negative, false rumors about you, or say negative, and ridiculing comments to your face. They may cause physical harm to you. An enemy will never help you when you have trouble. An enemy will demoralize a person just by constantly nit-picking them. Now that we have a few actions an enemy will perform, it becomes very easy to define a friend – it's the opposite of those actions. A friend will say kind things about you, spread good thoughts, and tell of your good deeds to others, they will be positive to you. A friend will encourage you, and motivate you. If you need physical care, a friend will help you physically, they help you to heal. When you have problems of any kind, a friend will give advice, or do things to help your situation.

So often, a person in our lives does so many 'enemy deeds,' yet we call them a "friend." They are not. There are a few reasons for this mislabeling of who is actually a friend. First, people don't make all the right friend moves, one hundred percent of the time. A true friend may do "enemy-actions" once in a while. They may regret their action and apologize. Still, it is easy to see who is an enemy, and who is a friend, by what the person's actions *predominantly* fall under. Does this "friend" perform mostly enemy actions? Or mostly friend actions?

The next reason we mislabel who is a friend is that we are taught, and expected to, be friends with certain people. For example, a family member, or a well-known person that is supposedly cogent in some desirable field. Isn't a parent *supposed* to build you up? Take care of you in times of trouble? Isn't a parent *expected* to take care of your physical needs? Treat you kindly? So, most people are expected to love and have friendship with their parent(s), or sibling, or cousin, or a childhood friend. These could all be lumped into the group of people that we are *supposed* to, we are *expected* to, be close with, and have a friend relationship. The truth is, some of these "closest-expected-friends" in our life are (in some cases) some of our worst enemies. We sometimes mislabel who is a friend because we think a person is "cool" or, they have some desirable quality that we admire. These people that we put on a pedestal, we long for them to be our friend. Sometimes this talented, popular, or "cool" person, is really just a boor, and not a friend at all. Family members and people we admire are not always friends.

One friend quality that seems to have a sliding scale of importance, is having similarities in beliefs, to have similar ideas and opinions. In today's culture, on social media platforms, it is common for people to state some strong belief they have, and then state something to the effect of, "If you don't believe exactly like me, unfriend me now!" This begs the question, is it possible to be friends with those who have different beliefs? Even on hot-button issues such as: God, or guns, environmental issues, what we eat, abortion, war, sexual relations, economic strategy, anything, any priorities about what is important in life? Can people with differing beliefs still be friends? It is easy to see that it is less stressful, there would be less worry about offending, if you are friends with people who believe the same way as we ourselves do. However, do not these people that disagree with us on any given issue actually make us grow? Are friends who have differing beliefs from our own, possibly, a *better* friend, because they help us to see life in a different way? How about if the person who disagrees with us on one "big" issue, say, whether we should be a vegetarian, or not, yet this person does all the other actions that a very close friend would? This other person builds us up, helps us, is kind to us, we delight in each other when we see each other, but we just disagree on whether eating meat is right or not. It is probably true that some beliefs may carry more weight, and these 'non-negotiable' beliefs are more important to agree upon in order to have a friend relationship. However, being friends with people who disagree with us, is a valuable tool for growth. One thing is for sure, if we only have friendships with people who belief exactly as we do, we will never progress to new, and possibly superior ways of living. The other important idea to consider is, if you cut off friends because of differing beliefs, or they disregard you, your beliefs and ideas about life may change over time. The bolstering of current beliefs by friends, could in many cases, be merely a transient occurrence.

Chapter 24

Ubiquitous Continuum #4: Magnanimity/Selfishness

Altruism, the value of being selfless, having concern for others, being considerate, this value has high impact on all our thoughts and life decisions. Of course, the opposite of this value would be selfishness. This value continuum between selfish and magnanimity has massive implications on the previously discussed, ever-present, destructive - constructive continuum. Every decision we make, whether it be to build benevolently, or destroy mightily, could all be placed along the continuum as to how selfish we decide to be at any given moment. Is it right, or good, to consider others' needs before your own? Is it *always* right to consider others' needs before your own? In reality, if we are honest, it certainly feels good to be selfish at times. Sometimes, to get things to "go your way," well, that often feels great! Even the decision to be selfish is as changeable as our decisions about what is right, and what is wrong, since each situation changes our choice in how to act.

Overall, it seems most humans value constructive behavior to help in living in close proximity to each other. Constructive behavior, selfless living, another synonym would simply be, kindness, helps us humans to get along. Here's an impossible future, everyone treats each other with constructive kindness… all the time. This hypothetical "impossible experiment" would be that everyone, on the entire earth, at the same time, decided to treat everyone else with dignity and kindness. Is it even imaginable? Let's hypothetically say, that everyone on the entire earth decided to make being a "constructive/kind person" the law. Let's say we all will, all at once, be kind, loving and respectful to each other, all the time. We all agree! We all do it; every human on the planet decides to treat each other, always, with respect and dignity. This fits many peoples' ideas of a utopian wish for our world. However, there are obvious problems with this 100% "world of kindness." Not only the fact that not all people would follow this, but it is not possible, because a kind act to one person, can be, and certainly will be, on occasion, perceived as a destructive act to another person.

Even a performing a "good/kind deed" of loving kindness may infuriate somebody else. There are multitudes of examples. Let's take dating for an example. Say you feel affection toward a person; you ask the person on a date. By doing so, you just hurt the feelings of the other person who is smitten with the same person you just asked to go on a date. You may not have intended to harm the other person's feelings, but you did. This simple example shows that kind actions can be, sometimes, unknowingly destructive and hurtful. Some people enjoy giving other people genuine compliments. Imagine the scenario where someone gives a woman a compliment on her pretty, long, brown hair. There happens to be another girl standing next to the complimented woman, she gets her feelings hurt that she didn't get the compliment. Here is one more scenario of kind acts that may be viewed as unkind. Say I'm in traffic, stuck in a long line of backed up cars. Here comes a car that needs to merge in, I, in my kindness, and respect for the merging car, motion for them to cut in front of me. The person behind me is aggravated with my kind action of letting the merging car in, for now, they are delayed even further. There are so many real-world scenarios where constructive/kind actions can often be viewed as destructive and rude to the viewpoint of another person. These are just simple examples to show that even when acting with intentions of constructive kindness, people can perceive disrespect, ridicule and/or frustration. Kindness cannot exist without destructiveness. Just as everything in this existence, our very definitions and understandings are linked directly to our perceptions on the continuum of opposites, in this case the "kindness – malevolence" continuum.

Here is another hypothetical, but real-world situation, to illustrate universal kindness is not possible. Imagine yourself working in a job and you are an efficient, and excellent worker. As an employee, you are a kind and worthy asset to your boss by being an outstanding worker. Unpredictably, by working diligently, you may have made someone else (The less industrious worker) feel poorly about themselves. You made them feel jealous, and possibly inferior, for not being as good of a worker as you are. Your "constructive" work ethic could be perceived as an act of unkind, selfishness to the person who is lazy.

All of the hypothetical situations illustrated above, show the impossibility of a perfectly kind and respectful world. In addition, there is always an individual prioritization of values. In an overall sense, there are values and morals (virtues) that are better to live by (the morals that help living in safety) than other morals and values. This, consequently, leads us back to the ever-present question of right and wrong. Prioritization of values leads to perceived injustice and disrespect. For example, if I say hard work is better than laziness, I am disrespecting the lazy person. If I say building other people up with compliments, is a good way to live, I am disagreeing with the sarcastic person, who enjoys ridiculing others. If I say putting other people's needs before your own is the right way to live, I am downgrading the ways of the selfish person. Who is kind, selfless, and hard working all the time? Nobody. These ideas can be applied to social issues and/or religious beliefs, on subjects such as sexual orientation, modesty of how we dress, and what we eat or drink. Some people are disgusted by, or disapprove of, other people's choices as to what is a good priority by which to live. The person making judgment on some action being good/right/better than another, wants to correct and teach/indoctrinate/coerce that other person. Entire wars come from this very fact. (Capitalism versus Fascism, Catholicism versus Protestantism, Muslim versus Christian, etc.) Our powerful culture's subjective choice of some superior behavior can be absolutely deadly to others. Two foundational aspects to this realm are always occurring: being magnanimous/constructive and being selfish/destructive.

Chapter 25

Religions and Governments Deter Freedom

Is there a better way to govern groups of people? Is there a more beneficial way for people to exist? Fascism? Communism? Socialism? Anarchism? Authoritarianism? Theocracy? etc. Is there a superior way to have a more fulfilled life under a certain type of governmental leadership? The role and actions of government would change according to differing opinions. Is government there to ensure a level of freedom and safety, that is most beneficial to the greatest amount of people? No one is ever truly, completely free, but to what degree could individuals be free under the different types of governmental rule? Since right and wrong is moveable for each society, (and even changes in individual lives throughout a single day) it is likely that *no* central government, or ruling religious body, should force their morals and priorities on other people. A person would benefit and experience a more fulfilled life, if they could make the decisions *they* wish to make without central law, and government interference. Of course, there are many people who choose to be spoon-fed someone else's ideas, accept these ideas as truth, and devote their lives to those ideas.

There is a sense of safety in narrowing your choices and "accepting by faith" someone's else's certainty in beliefs from big religions, a friend, or parents, or an advertisement, your government, or whatever.

Wouldn't it be wonderful if you could do your own research on any given subject, and then decide for yourself what action to take? After coming to any decision about how to live, it would be beneficial to *not* have interference from governmental law, or religious rules. Let us examine the subject of sugar intake, because in my current society, there are spurious views about the consumption of sugar. Spurious laws come from spurious beliefs. The vast majority of people do not know what constitutes sugar in any diet, or how the human body processes sugar. Yet, very few people would say sugar is *healthy* and *great* for your body in my current society. The truth is, **too much** sugar intake, that is what is harmful to the body. If we look at the science behind metabolism and carbohydrate intake, it is clear that the current, minority view that sugar, in moderation, is healthy, is actually the correct view.

First, let us tackle why sugar is good, then the detrimental effects of a misinformed leadership. All sugars are forms of carbohydrate, their primary role in the body is to produce cellular energy. Who wants energy? Who wants improved cognitive brain function? Who wants improved memory? Who wants reduced anxiety? The intake of sugars/carbs, has all of these documented, massive benefits (Ginieis et al., 2018; Berk et al., 2018; Haskell-Ramsay et al., 2017; Tufts Health Letter, 2013; Starkey, 2008). There are three forms of sugars: 1. Monosaccharides, which are simple sugars including glucose, fructose and galactose. In our body, fructose is converted into glucose by the liver then released into the blood for energy. The second form of sugars are 2. oligosaccharides, which is the sugar in the fiber of plants. This form of sugar is made of short chains of monosaccharides linked together. This is found naturally in beans, cabbage, and grains. The third, and final type of sugar, are 3. polysaccharides, these are long chains of monosaccharides linked together. These are foods with starch and cellulose such as corn, potatoes, bread, rice and pasta (Matthews, 2014).

All three of these types of sugars (monosaccharides, oligosaccharides and polysaccharides), **ALL** of them, end up as glucose in our blood. After our carbs are broken down into glucose, our body cannot determine the difference between honey, peas, potatoes, oranges, milk, pasta, or a candy bar… they are all sugar. All sugars are broken down to monosaccharides in our blood, namely, the monosaccharide called glucose (Matthews, 2014). It is true that some foods are broken down by our bodies into glucose at a faster rate than others, but that is the only difference between any carbohydrate we put in our mouths. Simple sugars (Monosaccharides) elevate insulin levels higher, and faster, than polysaccharides found in vegies and pasta, but still, remember, all carbs are broken down to glucose. Regular old, classic, white table sugar, is one-part fructose and one-part glucose. Sucrose is found naturally in many fruits, sugar cane, and nuts (Zelman, 2024).

Ask someone who has type 1 diabetes, I have. He told me that if he needs to raise his sugar level, the quickest, "sugar-iest" way, is a spoonful of honey under his tongue. This carbohydrate makes a very rapid increase of glucose in the blood. Why is there no outcry against honey? It is probably because we, as a culture, have decided honey is natural and quite healthy, and it is. M&Ms raise blood sugar as well, not as quick as honey, but within about five minutes, blood sugar rises. This same friend with diabetes also told me that pasta, potatoes, and starches, they also raise blood sugar, but it takes longer. Plus, importantly, there are also many other factors determining sugar level in our blood: stress, anxiety, the amount of physical activity we perform, sugar through liquids, exercise, and which foods are paired with each other… all of these factors, on an individual level, have impact on how sugar is used in the body. If an off-the-charts, sugar-spiking pomegranate is eaten, right after some protein, it will not increase sugar level in the blood as quickly or severely (Loneman, 2024; Ryan, 2024). The point is, when we know, at the molecular level, how our bodies breakdown the three basic types of sugar, the ignorant laws that limit sugar intake would never be put into place.

Government leaders come along and make faulty rules and laws for *everyone*, based on their twisted, biased, influenced-by-culture "science," and cultural pressure. Propaganda and band wagon followers create a tide of idiocy in which all of us get swept along. Schools ban drinks, cupcakes, and candy. The city of New York, banned extra-large soft drinks, all in the vein of forcing the unlearned rabble to do the "right" thing, and stop eating sugar. The ban on large soft drinks in New York City was eventually overturned (International Business, 2012; Klepper & Peltz, 2014). The entire country of Singapore banned soft drinks (Cheung, 2019). Accordingly, when rules are made, that's when people are stripped of freedom of making their own choices. These edicts, whether beneficial, or not, whether they are false, or not, stop creativity, and in the case of sugar - health. These laws and rules hinder, or halt, "outside the box" ideas that might actually make improvements to this life.

The bottom line, sugar, in moderation, is truly healthy, and great for your body (Ginieis et al., 2018; Berk et al., 2018; Haskell-Ramsay et al., 2017; Tufts Health Letter, 2013; Starkey, 2008). It is when we overindulge in sugar, that is when it is bad for you. There will be some people, who read this evidence that sugar, in moderation, is healthy. They will still choose to believe sugar is bad. Similar to challenging someone's religious/faith beliefs, often times, even after confronted with science, or straight-forward rational ideas that challenge their deeply entrenched beliefs, they will still stay with their comfortable, false belief. There is a comfort in consistency. Also, there is the issue of pride, it's difficult to admit to self, or others, "I may have been wrong."

Especially if a belief has been bolstered since early childhood, and all your surrounding culture still believes the falsehood. In any case, deeply held beliefs by many, can stop the freedoms of others.

You might be saying to yourself that this is a small thing - who cares about sugar? You might even still falsely believe that a candy bar and a potato are totally different, that one is harmful to the human body, and other is not. Regardless, there are a multitude of questionable laws, rules and statutes that stamp all over our rights, and our ability to truly be free in the choices we make. Arguments about immigration, smoking, guns, homosexuality, animal rights, fluoride in your drinking water, standardized testing, taxes, vaccinations, seat belts, alcohol, marijuana, abortion, just to name a *few* topics off the top of my head. Choice is taken away when laws are made. Laws are enforced to encourage some behavior, or to suppress peoples' behavior. This stops progression into new and better ideas about how to live.

My purpose here is not to argue about which side of any of the listed topics is right or wrong. Obviously, from the entire first part of this book, right and wrong changes. I am trying to point out the enforced tunnel vision that blocks beneficial progress. So, why do we have rules? Why does the government (even under democratic/representative governments) see fit to lord their chosen ideas, and ideals, on everyone else? We can guess motivations of possessing dominating power. For some people, who deem having power is a virtue to live by, forcing others to accept their view on any subject is a pleasurable, and a worthy goal for some. It is certainly possible that the government, or ecclesiastical rulers, really do believe they have everyone's best interest in mind. Churches, groups, and clubs have the same function as civil governments, to spread their beliefs to everyone else. In recent times, here in the United States, in many cases, these religious groups and/or clubs don't have *legal* right to enforce their priorities. Churches exert pressure in other ways, through shaming, or access to higher levels within the group. The problem is, limiting choice creates tunnel vision, as the enforced thinking *is the only* way to go.

Allegedly, most rules and laws are made to ensure safety, but what if we had individual freedom with one overarching rule: treat each other with respect and kindness? What if everyone was free to do whatever they wanted, as long as it didn't harm others? The word anarchy is scary, it has connotations of craziness and lawless evil, but anarchy does not have to be crazy and hedonistic. People existing without governmental rule can occur if everyone had everyone else's best interest in mind. "Just care for one another," sounds so hippie-ish. Care to choose actions that build others up, as much as possible. Building others could occur in a physical way, a mental way, or an emotional way. If all humans were truly empathetic, kind, and continually *trying* to complete actions that helped other people, this would be as close to utopia as humans could get. As I have covered earlier in this book, complete kindness could never happen due to the varied viewpoints on any single action. As with all governmental systems, or the lack there of (anarchy) coupled with treating others with kindness and respect, there are obstacles. The problem isn't in having no central rule, it could work. The problem lies in the enforcement, and perception of what constitutes an act of kindness. Enforcement of love in an anarchy, is incompatible (making the virtue of love a law), but humanity should at least strive for this, so individual thoughts and ideas can bloom. Think of the progress that humanity could make if we weren't limited by religious, or governmental rules. The approach to life without government would have to have people living by the maxim of building others up. This is obvious when you consider the opposite end of that continuum: building-others-up on one side, and on the other end of that continuum would be… destroying each other.

How to deal with individuals (or groups) that condone killing others is contradictory. One example, and I could pick many examples, but in the United States, in the year 2025, it's wrong to murder people that don't have the same beliefs as yourself. The overwhelming majority of rational people, currently believe that it is wrong to kill another person because of their beliefs. Yet, we have had the Twin Towers bombed in New York City. There was another incident in April of 2013, when bombers from Turkey, blew up people during the Boston Marathon (Boston Marathon, 2012). Just recently, in Paris, France, and Las Vegas, Nevada, shooters killed innocent people in concerts. The response is always immediate revenge. State law enforcement officers mobilized to kill the people that murdered others for having different beliefs than their own. Therein lies the contradiction, we have this societal/legal agreement that murder shouldn't occur due to our beliefs, yet we deliberately go, and murder the murderers. Paradoxically, enforcement, of even pacifist tenets such as kindness, compassion, and love, are, at times, done with murderous force.

A straight forward example of killing over beliefs is to look at wars of the past. In the 1960's and '70's the United States government believed capitalism is such a vastly better governmental system, in order to stop Vietnam from becoming a communist country, *millions* of people were killed. Nearly 60,000 U.S. soldiers' deaths comprised a part of those millions that died (National Archives, 2008). From the 11th to the 13th century, Islamic and Christian believers killed each other in the millions, because each side thought they had the correct moral/religious beliefs, in their correct God. Killing over beliefs has happened all throughout history. Here are just a few:

-Israel/Palestine has done this since the 1970's

-Bosnia in 1990's killing between Bosnian Muslims
 Orthodox Serbs, and Catholic Croats

-The U.S.A. in the 1860's and 1870's beliefs on slavery, and native
 inferiority.

-Germany in the 1940's

-Japan in the 1930's

… all attempted to make a "superior" society by deciding that certain groups with differing beliefs, should be exterminated. "Cleansing" the society of what some person, or people, decided was inferior, destructive, or wrong. Governments and powerful religious groups have this terrible, murderous ability toward others that disagree with their beliefs, or their vested interests. How many times throughout all of history, right up to this present day, have plebs died because of their *leaders,* "superior" beliefs?

Political theorist, Carl Schmitt, put forth that one of the supposed purposes of using a democratic republic (with representatives of the people) for governmental rule, was to get rid of the sovereign rule, by an all-powerful individual, such as a king, or a president. Ostensibly, a representative government's benefit is to give more say to a greater number of people from differing social, and economic levels of society (Busk, 2012). Groups of humans (countries) have periods of regular, peaceable, functioning, during these times there is no need for control by religious or governmental leaders. Then, there are other times, when a leadership group with monetary, or military power, want a desired change. When this change, or manufactured "crisis" comes along, it becomes readily observable that democratic/representative governments are a sham (Vinx, 2021; Zielinski, 2022). When the Covid pandemic came along in 2020, all the wealthy, true ruling elite, who are in control, were easily visible. The leaders *demand* what happens. Vaccination shots were mandated; people were not even allowed to leave their house. Businesses were closed down. Rights and laws give the deception that the world is a more tolerant, kind, and forward-thinking place of freedom… but it's not. The wealthy, influential monarch, leader, or oligarchal leadership group, still rules the average, everyday person.

Pondering Our Existence for Practical Living

Permits to work construction on your own house, gun use, freedom of speech, helmets, what food or drink you consume, worship decisions… all of these issues and many, many more, show a ruling class never stops hampering freedom of choice. It sounds like a worthy goal, to have humanity flourish in democratic societies. Government by the many, to listen to, debate, and settle on new ideas. The problem is, if a person doesn't accept the current, majority-approved ideas, well, that person's approach is done away with, whether you are in a representative government, or not. Political agendas so often devolve into coercion, by force, to mandate a point of view, or course of action. This happens in every style of government, or leadership group. Even in a perfectly working democratic government, the majority can, and often does, overrule the minority; the majority's will *is* the will of the people. A representative system forces a distinction between people who have governmental representation, and those who do not agree with their elected representative(s), and thus, these people are not represented. Even if a true democracy was in place, the rule of the majority will always have a form of indirect rule of one social group over another. The possibly superior views of a minority, will be suppressed.

A brief look at all of history shows that a unified group/country is able to wield more military power, and can take over fragmented societies. This can be viewed as good, if imperialism is accepted as a worthy goal. A technologically superior, and/or united force, will subjugate, or annihilate another group. In dictatorships, or in governments where the citizens are completely controlled by the state, those groups are unified, ordered, and decisive (Wintrobe, 2001). There are many examples throughout history: Macedonian Military force, or Roman, Frankish, Lombard, Dutch, German, English, American, Russian, or any imperialist power from of the past, and present, often dominates the areas with less unified groups. Conquistadors destroyed the Aztecs, the USA destroyed the Native American tribes, Romans destroyed the disjointed Celts, Norse Viking raiders destroyed Irish tribes, the Normans took over fragmented Southern Italy and the British Isles. This could be accomplished because the groups they dominated were smaller, technologically inferior in warfare, and divided. Imperialism still occurs, but the conquering imperial nations and countries are supposed to feel horrible about their past aggressive deeds. An anarchic society, free of government rule, is inferior in the ability to ward off imperialist groups that are in lock-step with leader mandates. This is obvious even without all the above examples. In a group of free individuals, some would certainly decide *not* to fight for their group. In a group where behavior is forced, most people under central government rule, would be in lock-step with their group. Imperialist dominance shows allied groups/societies are more powerful, militarily, than solo families, and unaligned individuals arranged in fragmented communities.

Socially accepted ways of how to behave are defined by such ideas as rights, freedom and opportunity. Yet, inequalities exist within any government-led societies. The style of governmental rule does not matter: communist, socialist, capitalist, dictatorship, etc. The existence of inequality is inevitable. People are born with different degrees of ability in different areas. People are nurtured in one specialty area, or another, by their parents. Some inequalities are based on some unfair advantage. Other inequalities come about due to someone's effort. Sometimes, it is difficult to distinguish between the two. Inequalities are not limited to wealth only… people are continually born into unequal/unjust situations. There are pervasive, observable, situations of inequality: Intelligence, mental illness, beauty, height, abusive family situations, charisma, athletic ability, horrible resource areas, low emphasis on learning, high crime areas, etc. We are born into a genetic and cultural situation (whether it's viewed as fair, or unfair) that has a massive impact on our existence. How do you stop a person from being more intelligent than another person? (Intelligence can be defined in various ways.) How do you stop someone from being more beautiful than another? (Beauty can be defined in various ways.) How do you stop unequal and varied family methods of raising children? (Successful child rearing can be defined in various ways.) Are not inequalities acceptable if the inequalities come from hard work, or extreme effort? Inequality, considered just, or unjust, is impossible to escape. Consider a person who has a natural advantage in athletic ability, good looks, or has musical talent, these gifts are beneficial for all of society. Entertainment is one of the joys of this life, do great actors have an unfair advantage? The extremely intelligent person, does have an unequal/unfair advantage as a "thought leader," but this person raises the bar for all of society. Which inherited inequalities should be labeled fair, or unfair? A government that limits its top people, hurts all of society. Examples would be Galileo being censured for his scientific discoveries. Also, economic interests that stop fuel/energy alternatives that would help all. If the superior, or the advantaged, causes harm to someone else, that is detrimental. An effective, progressing society, should enrich each other with whatever skill they have. Negative, uncontrolled circumstances should never impede someone's ability to live a fulfilled life, but hard work and perseverance should never be stopped either. The state of nature itself has elements of wanton destruction as well as moments of care. Is it possible for humans to break with nature and structure a more peaceful society? What leadership situation helps the least advantaged within that society, in comparison to all others? What societal structure offers the least advantaged person in society a better chance for a fulfilled existence?

Justice is fairness, not equality. It would be advantageous to do away with social stratification, a great life (however one measures that) should be of the making of each individual. Fairness in society would be beneficial if fairness focused on opening economic gain (or basic physical needs stability) for everyone. If after a person exerts effort, earns through hard work, and there are differences in quality of life, this should be acceptable. If there is a massive discrepancy between the quality of life, the advantaged should never be allowed to hinder to the disadvantaged. Social and economic inequalities must not affect positions that should be open to all. Equality of opportunity is paramount. The inequality is acceptable if society has truly made access to all a reality. Why stop a thoroughbred race horse that wants to sprint? Take off the bridle, and let the thoroughbreds of society run! Stopping a top academic, a top athlete, or a hard-working person, now that would be truly be unfair. Just societies would allow inequalities in the areas of talent and effort. Hopefully, every person has a voice, their ideas are able to shine forth. The encouragement of people to become as talented as they possibly can, this will raise all of society.

If a person kills another person by purely accidental circumstance, the "killer" would rarely be sent to jail (or have a far less punitive sentence) for murder, because the killer isn't a "bad" person. They merely were a part of a non-malicious mistake. In the same way, your life should not be penalized (or rewarded) for random, unfortunate/fortunate circumstances. For example, not being born into a wealthy family. People are not good, or bad, because they are born into wealthy circumstances, just as they are not good or bad due to a random accident. It is important to recognize our innate longing for equality.

Is it a correct way to live to have billionaires, existing in the same society where there is hunger and extreme poverty? Very wealthy people and very poor people are living simultaneously under the same governments. Some would say this is an unfair, or even an "immoral" setup. Some would say this is an acceptable outcome because these two types of people (poor and rich) have made different life choices, and they have taken different actions to get to where they are in life. The lucky circumstances of some being born into wealthy families can't be stopped. However, how many hours a person works, and what they do to make this money, is based on individual effort. Income is an ineffectual measure of equality (or inequality) because the difference is in how different people use their own liberty of running their own lives. What justifies inequality? What if the billionaire got his/her money through robbery? What if the billionaire earned it? What if the billionaire inherited it? Inequality seems to be acceptable, as long as it is based on someone's individual work, or effort, that they have put forth. Importantly, individual effort and talent, is most able to flourish when a government is not stopping people from doing so.

Stepping back from the distribution, fair use of money, and government tactics, the belief and honor given to any monetary system by any society is truly bizarre. Dwell on this for a moment, humans decide a certain paper, or coin, or mineral, or (now) just a number in a bank account, means something. Humans could easily choose virtually any object, manufactured, or natural, and all agree, "this (fill in the blank) is the object that has value." It is completely arbitrary, yet we're all slaves to this chosen currency. Now, people that have lots of this decided-upon-currency, have a lot more freedom to do what they want to do, when they want to do it. People that don't have the decided-upon-valuable-currency, have less freedom to do what they would want. If the people with lots of wealth can innovate technologies, and give philanthropically, this could be an inequality that *could* turn to everyone's advantage. Being a rich person does not have to be a bad thing. Through our own individual skill sets, we all should help each other. Plus, the advantaged people only got to their status with help from all. The intelligent person, born into a nurturing family, access to learning, safe neighborhood, lucky to remain healthy… this person only got to their position because others have made this possible. The people who grew food for this person, made roads, paid for the schooling, gave words of encouragement, built the school, they all made this advantaged person's life possible. This is reason enough to ensure the rich are not too rich, and the poor are not too poor, all people are integral to each other's advancement.

There are, and have been throughout history, overbearing government systems. Let's use the fascist, Nazi government in Germany during the mid-twentieth century. The young were taught that giving your time in this life for the current state (Germany) in which they lived, this was the greatest good, and fulfillment of life. To devote your time to the leaders, or leader, of the country into which you have happened to be born, this has been the indoctrination young people were taught (Aburto & Beltrán-Sánchez, 2019). To devote your life to your ruling government is paramount. This seems nonsensical to many people. The governments of the many countries around this globe are so different. The governments of the Congo, the United States, Russia, England… anywhere on the planet, have different versions of governing. Their version of good or evil is based on opinion. What makes living for the Italian government better, or worse, than living our life to fulfill the longings and needs of the Philippine government? Why complete actions to support any government? It is utterly senseless how many people have died throughout history because of some leader/government's cause. Governments are just made up of people (or a single person) anyway, and these leaders are going to decide for you what priorities are right? What is good? What is fulfilling? This is the same situation where people adopt what others tell them in the realm of religion, diet, morals, whatever. Say my current governmental overlords in the USA decide they want more (and cheaper) oil. They decide on war to procure this oil. What if that chosen priority does not coincide with my personal beliefs? I am then trapped and mandated to be a part of actions which I really do not want to be a part of! This is an important reason as to why having no central ruling body is advantageous. For some people, the opinion that living one's life for their government, well, would be the least desirable way to waste your existence.

It seems beneficial to have the freedom to choose your own values, your own actions. Will your own freedom and projects, get in the way of the freedom of others? Will they be stopped by some governmental rule? True freedom requires the freedom of others, because we are a member of our surrounding society. Maximize the freedom of people, then, maybe, we will find our way to new understanding. A new way to a better existence.

Poisoned and wrecked, are we bent toward wreckage?
Do we know what we want?
What we're getting into at the time?
It's like a river, impossible to buck.
It takes you downstream
You can't help the wreckage and decay.
Pristine to decay.
Beautiful to unpleasant.
Unpleasant to dead.
Don't poison me
Don't poison yourself
Wait, that means don't live
We defile, we can't help it.
We can't help but to be defiled.

Are you coming back?
Let's get it right.
You get it right?
Will you?
Is that stupid?
Who's coming back?
Why perfect it?

Pondering Our Existence for Practical Living

I can't sleep because there is pain in my heart
I can't sleep because there's pain in the world
I can't sleep because I have pressure
Jobs, people, chores to be done
And all the while embarrassment and death loom close by.
I can't sink to sleep because my mind is wide awake
The waves of care have crashed on top of me
I can't rest,
I'll drown if I stop treading

What a life
We're thrust into a scenario as varied as snowflakes.
We're coached, or not, in how to live in this time and space, while bumping into other souls.
Interactions are painful and beautiful and teach us anew.
Although, the fog and haze still hangs heavy and cloud our vision of who we are and what we need to be;
I don't think we ever get a glimpse beyond the next candle-lit step.
One day, eternity comes, and everyone plans for a setting of the record straight.
Where did they go?
Will I know you again someday?

Stephen Gentile

Have great dreams
Take the world for a ride
Talent never to hide
Someone said you can't
With a huff and a pant
You believed them and left
Rebel against the world
Leave its ways behind
Live life for yourself
You're a legend in your own mind
Fate takes you through
You're thinking you've thought it through
Having grand ideas in mind
Maybe someday they'll discover
What a special soul I am
All the talent… It's not just a sham?
Remembered after the fact of living
Like Shem, Japeth and Ham
You'll be the joy of some land
You're only a bit out of hand
You're a legend in your own mind

Pondering Our Existence for Practical Living

I'm something – I'm impacting
I'm exacting a strong response
Be it a small venue
I'm orchestrating my personal renaissance
I'm reeling against an enemy
An enemy within, the one that tells me I'm everything
When in reality I can't begin to approach the insecurity of it all
Do I really have the gall? To say I'm not so small?
I'm a legend alright, a legend in my own mind.
My offspring thrill me daily
Daily I'm reminded; I may not be somebody whose life is worth rewinding,
But I'm floored at the movement and the being of my child
The feelings are so wild, as a fox whose been beguiled
I'm rewinding it myself to breathe in the joy and painful, jubilant mime
I do believe I'm a legend in my own mind.

Bibliography

Ataei, A., Amini, A., & Ghazizadeh, A. (2024). Robust memory of face moral values is encoded in the human caudate tail: a simultaneous EEG-fMRI study. *Scientific Reports, 14*(1). https://doi-org.lili.idm.oclc.org/10.1038/s41598-024-63085-w

Armbrüster, T. (2005). Management and organization in Germany. Ashgate.

Antoniadis, I. (2015). Quarks and a unified theory of Nature fundamental forces. *International Journal of Modern Physics A, 30*(2). https://doi-org.lili.idm.oclc.org/10.1142/S0217751X1530015X

Aburto, J. M., & Beltrán-Sánchez, H. (2019). Upsurge of Homicides and Its Impact on Life Expectancy and Life Span Inequality in Mexico, 2005–2015. American Journal of Public Health, 109(3), 483–489. https://doi-org.lili.idm.oclc.org/10.2105/AJPH.2018.304878

Alves Vicente, G., & Antônio Witt, M. (2018). A Educação Na Alemanha Durante O Terceiro ReichE Seu Papel Na Doutrinação Das Crianças E Jovens. *Conhecimento Online, 10*(1), 71–87. https://doi-org.lili.idm.oclc.org/10.25112/rco.v1i0.1179

Reference List

Anderson, J., & Holland, E. (2015). The legacy of medicalising "homosexuality": A discussion on the historical effects of non-heterosexual diagnostic classifications. *Sensoria: A Journal of Mind, Brain & Culture, 11*(1), 4–15. https://doi-org.lili.idm.oclc.org/10.7790/sa.v11i1.405

Aristotle. (2012). *Aristotle's Nicomachean ethics*. University of Chicago Press.

Arkani-Hamed, N., Langer, C., Srikant, A. Y., & Trnka, J. (2019). Deep Into the Amplituhedron: Amplitude Singularities at All Loops and Legs. *Physical Review Letters, 122*(5), 051601. https://doi-rg.lili.idm.oclc.org/10.1103/PhysRevLett.122.051601

Arkani-Hamed, N. (2012). The Future of Fundamental Physics. *Daedalus, 141*(3), 53–66. https://doi-org.lili.idm.oclc.org/10.1162/DAED_a_00161

Atman. (2018). *Funk & Wagnalls New World Encyclopedia*, 1.

Baptism as a Heathen Rite. (1881). *Nation, 33*(847), 238–239.

Baruah, S. (2019). Understanding the Philosophy of Buddhism and its challenge to Brahmanism. *Clarion: International Multidisciplinary Journal, 8*(2), 52–55. https://doi-org.lili.idm.oclc.org/10.5958/2277-937X.2019.00018.2

Batygin, Konstantin (2008). "On the dynamical stability of the Solar system". *The Astrophysical Journal*. **683** (2): 1207 1216. arXiv:0804.1946. Bibcode:2008ApJ...683.1207B. doi:10.1086/589232. S2CID 5999697.

Reference List

Beauchamp, M. S., Oswalt, D., Sun, P., Foster, B. L., Magnotti, J. F., Niketeghad, S., Pouratian, N., Bosking, W. H., & Yoshor, D. (2020). Dynamic Stimulation of Visual Cortex Produces Form Vision in Sighted and Blind Humans. *Cell*, *181*(4), 774.

Belgum, D. (2023). Robberies In L.A. Leave Retailers Unsettled. *WWD: Women's Wear Daily*, 1–14.

Berk, L., Miller, J., Bruhjell, K., Dhuri, S., Patel, K., Lohman, E., Bains, G., & Berk, R. (2018). Dark chocolate (70% organic cacao) increases acute and chronic EEG power spectral density (µV2) response of gamma frequency (25–40 Hz) for brain health: enhancement of neuroplasticity, neural synchrony, cognitive processing, learning, memory, recall, and mindfulness meditation. The FASEB Journal, 32(Supplement 1), 878.10. https://doi-org.lili.idm.oclc.org/10.1096/fasebj.2018.32.1_supplement.878.10

Berkowitz, R. (2024). Flowers might gossip via electric fields. *Science News, 205* (8), 12.

Bicudo, P., Cardoso, N., & Sharifian, A. (2022). Highly excited pure gauge SU(3) flux tubes. *EPJ Web of Conferences*, *258*, 1–7. https://doi-org.lili.idm.oclc.org/10.1051/epjconf/202225810001

Bierschbach, B. (2012). Minnesota pro-marriage amendment group releases two television ads. *St. Paul Legal Ledger (MN)*.

Reference List

Blackledge, J. T., & Hayes, S. C. (2001). Emotion Regulation in Acceptance and Commitment Therapy. *Journal of Clinical Psychology, 57*(2), 243–255. https://doi-org.lili.idm.oclc.org/10.1002/1097-4679(200102)57:2<243::AID-JCLP9>3.0.CO;2-X

Boehlke, C., Zierau, O., & Hannig, C. (2015). Salivary amylase - The enzyme of unspecialized euryphagous animals. *Archives of Oral Biology, 60*(8), 1162–1176. https://doi-org.lili.idm.oclc.org/10.1016/j.archoralbio.2015.05.008

Boston Marathon. (2021). Funk & Wagnalls New World Encyclopedia, 1.

Bower, B. (2013). Babies perk up to ancient threats: Snake's hiss, angry voices quickly capture infants' attention. Science News, 184(7), 11.

Bowling, N. C., & Banissy, M. J. (2017). Modulating vicarious tactile perception with transcranial electrical current stimulation. The European Journal of Neuroscience, 46(8), 2355–2364. https://doi-org.lili.idm.oclc.org/10.1111/ejn.13699

Brown, Dee (2001) [1970]. "War Comes to the Cheyenne". *Bury my heart at Wounded Knee*. Macmillan. pp. 86–87. ISBN 978-0-8050-6634-0.

Buenker, R. J. (2014). Application of relativity theory to the global positioning system. *Physics Essays, 27*(2), 253–258. https://doi-org.lili.idm.oclc.org/10.4006/0836-1398-27.2.253

Reference List

Buchanan, R. Angus (2024, October 23). *History of Technology. Encyclopedia Britannica.* https://www.britannica.com/technology/history-of-technology

Burger, B., & Wöllner, C. (2023). Drumming Action and Perception: How the Movements of a Professional Drummer Influence Experiences of Tempo, Time, and Expressivity. *Music & Science, 6*(1). https://doi-org.lili.idm.oclc.org/10.1177/20592043231186870

Burley, M. (2013). Retributive karma and the problem of blaming the victim. *International Journal for Philosophy of Religion, 74*(2), 149–165. https://doi-org.lili.idm.oclc.org/10.1007/s11153-012-9376-z

Burton, Kristen D. PhD (2020, July 16). "Destroyer of Worlds": The making of an Atomic Bomb. The national WWII Museum. Retrieved Feb 1, 2025. https://www.nationalww2museum.org/war/articles/making-the-atomic-bomb-trinity-test

Busk, L. A. (2021). Schmitt's democratic dialectic: On the limits of democracy as a value. *Philosophy & Social Criticism, 47*(6), 681701. https://doi-org.lili.idm.oclc.org/10.1177/0191453720916514

Carroll, S. M. (2019). Something deeply hidden: quantum worlds and the emergence of spacetime. E.P. Dutton.

Reference List

Chandrasegaran, J., & P., P. (2018). The Role of Self-Fulfilling Prophecies in Education: Teacher- Student Perceptions. *Journal on Educational Psychology, 12*(1), 8–18.

Cheung, E. (2019, Oct 11). Singapore to become first country banning ads on sugary drinks. *CNN Health.* Retrieved Feb 8, 2025. https://www.cnn.com/2019/10/11/health/singapore-sugar-drink-ads-intl-hnk-scli/index.html

Condon, L. (2000). Friend or Family? *Advocate, 823,* 48.

Cooney, E. C., Jacobson, D. M., Wolfe, G. V., Bright, K. J., Saldarriaga, J. F., Keeling, P. J., Leander,B. S., & Strom, S. L. (2024). Morphology, behavior, and phylogenomics of Oxytoxum lohmannii, Dinoflagellata. *Journal of Eukaryotic Microbiology, 71*(6). https://doi-org.lili.idm.oclc.org/10.1111/jeu.13050

Cope, S. (2006). The wisdom of yoga : a seeker's guide to extraordinary living. Bantam Books

Cox, J. (2024). *Women money power : the rise and fall of economic equality.* Harry N Abrams Inc.

Dale, D. (2012). U.S. Sniper's Book Claims 160 Kills Ex-Navy Seal Wishes He Could Have ShotMore Insurgents in Iraq. (2012). *Hamilton Spectator, The (ON).*

D'Ariano, G. (2017). Physics Without Physics. *International Journal of Theoretical Physics, 56*(1), 97–128. https://doi-org.lili.idm.oclc.org/10.1007/s10773-016-3172-y Einstein Confirmed. (1998). *Astronomy, 26*(3), 28.

Reference List

Daujeard, C., Falguères, C., Shao, Q., Geraads, D., Hublin, J.-J., Lefèvre, D., Graoui, M. E., Rué, M., Gallotti, R., Delvigne, V., Queffelec, A., Arous, E. B., Tombret, O., Mohib, A., & Raynal, J.- P. (2020). Earliest African evidence of carcass processing and consumption in cave at 700 ka, Casablanca, Morocco. *Scientific Reports, 10*(1), 4761. https://doi-org.lili.idm.oclc.org/10.1038/s41598-020-61580-4

Dialmy, A. (2010). Sexuality and Islam. The European Journal of Contraception and Reproductive Health Care, 15(3), 160–168. https://doi-org.lili.idm.oclc.org/10.3109/13625181003793339

Dias, E. (2014). Joseph Smith's Many Wives: The Faith at Stake in the News. *Time.Com*, N.PAG

Dicks, L. (2007). My family and other plants. *New Scientist*, 196(2635/2636), 64–65.

Duntley, M. (2005). Clergy, Discipline, and the Salem Witch-Hunt: Popular Stereotypes vs. 17th Century Ecclesiology. *Journal of Religion & Abuse, 7*(2), 57–68. https://doi-org.lili.idm.oclc.org/10.1300/J154v07n02_04

Duqum, André. (2024, Jul 23). The Emerging Science: "We Are ONE Consciousness" - Life, Death & The Simulation | Donald Hoffman. [Video]. YouTube. https://www.youtube.com/watch?v=ffgzkHCGZGE

Reference List

Fagan, Brian. (2004). The seventy great inventions of the ancient world. Thames & Hudson.

Farman, Henri. 1874-1958 Farman Chronology. *Early Birds of Avaiation.* Retrieved Jan 31, 2025 https://www.earlyaviators.com/efarman.htm.

Fesce, R. (2023). Imagination: The dawn of consciousness. *Physiology & Behavior, 259*(1). https://doi-org.lili.idm.oclc.org/10.1016/j.physbeh.2022.114035

Flaum, E., & Prakash, M. (2023). Curved crease origami and topological singularities at a cellular scale enable hyper-extensibility of *Lacrymaria olor . BioRxiv : The Preprint Server for Biology.* https://doi-org.lili.idm.oclc.org/10.1101/2023.08.04.551915

Franco, S., Galloni, D., Mariotti, A., & Trnka, J. (2015). Anatomy of the amplituhedron. *Journal of High Energy Physics, 2015*(3), 1–63. https://doi-org.lili.idm.oclc.org/10.1007/JHEP03(2015)128

Frater, Jamie. (2007). Top 10 Ancient Inventions. Retrieved Jan 31, 2025. https://listverse.com/2007/10/07/top-10-ancient-inventions/

Gabrielsen, P. (2013). Why Your Brain Loves That New Song. *Science Now,* 3.

Reference List

Gambino, M. (2009). A Salute to the Wheel. *Smithsonian Magazine.* Retrieved 1/31/2025.
https://www.smithsonianmag.com/science-nature/a-salute-to-the-wheel-31805121/

Garrison, P. (2003). *Wrightophilia. Flying, 130(12),* 89–91.

Gavrilets, S. (2012). On the evolutionary origins of the egalitarian syndrome. *Proceedings of the National Academy of Sciences of the United States of America, 109*(35), 14069–14074. https://doi-org.lili.idm.oclc.org/10.1073/pnas.1201718109

George, A. (2020). What you experience may not exist. Inside the strange truth of reality. *New Scientist, 245*(3267), 39–43. https://doi-org.lili.idm.oclc.org/10.1016/s0262-079(20)30222-0

Gillett, G., & Franz, E. (2016). Evolutionary neurology, responsive equilibrium, and the moral brain. *Consciousness and Cognition,* 45, 245–250. https://doi-org.lili.idm.oclc.org/10.1016/j.concog.2014.09.011

Ginieis, R., Franz, E. A., Oey, I., & Peng, M. (2018). The "sweet" effect: Comparative assessments of dietary sugars on cognitive performance. *Physiology & Behavior, 184*(1), 242–247. https://doi-org.lili.idm.oclc.org/10.1016/j.physbeh.2017.12.010

Global Slavery Index. (2023). Global Findings on Modern Slavery. *Walk Free.* Retrieved 3/14/2025. https://www.walkfree.org/global-slavery-index/findings/global-findings/

Godel Proves the Incompleteness of Formal Systems. (1999). In *Great Scientific Achievements* (Vol. 3, p. 253). Salem Press.

Goel, V., & Vartanian, O. (2011). Negative emotions can attenuate the influence of beliefs on logical reasoning. *Cognition & Emotion*, *25*(1), 121–131. https://doi-org.lili.idm.oclc.org/10.1080/02699931003593942

Gollee, H., Volosyak, I., McLachlan, A. J., Hunt, K. J., & Gräser, A. (2010). An SSVEP-Based Brain—Computer Interface for the Control of Functional Electrical Stimulation. IEEE Transactionson Biomedical Engineering, 57(8), 1847–1855. https://doi-org.lili.idm.oclc.org/10.1109/TBME.2010.2043432

Goldsmith, D. (2008). Dark Energy Crisis. *Natural History*, *117*(10), 30–34.

Gopnik, A. (2010). How Babies Think. *Scientific American*, *303*(1), 76–81. https://doi-org.lili.idm.oclc.org/10.1038/scientificamerican0710-76

Greene, B. (2004). The Fabric of the Cosmos: Space, Time, and the Texture of Reality (Book). *Booklist*, *100*(12), 1013.

Reference List

Greyson, B. (2021). After: a doctor explores what near-death experiences reveal about life and beyond. St Martins Pr.

Hadhazy, A. (2017). The Dark Universe. *Discover, 38*(6), 76–77.

Håland, E. (2009). Water Sources and the Sacred in Modern and Ancient Greece and Beyond. *Water History, 1*(2), 83–108. https://doi-org.lili.idm.oclc.org/10.1007/s12685-009-0008-1

Hameroff, S. (2021). "Orch OR" is the most complete, and most easily falsifiable theory of consciousness. *Cognitive Neuroscience, 12*(2), 74–76. https://doi-org.lili.idm.oclc.org/10.1080/17588928.2020.1839037

Hamper, D. (2024). Modern Slavery. *Legaldate, 36*(2), 1–6.

Hamzelou, J. (2022). A man communicates by thought alone. MIT Technology Review, 125(3), 10.

Harlow, H.F. (2018, Jun20). Harlow's Classic Studies Revealed the Importance of Maternal Contact. *Association for Psychological Science*. Retrieved 2/4/2025. https://www.psychologicalscience.org/publications/observer/obsonline/harlows-classic-studies-revealed-the-importance-of-maternal-contact.html

Reference List

Haskell-Ramsay, C., Stuart, R., Okello, E., & Watson, A. (2017). Cognitive and mood improvements following acute supplementation with purple grape juice in healthy young adults. *European Journal of Nutrition*, *56*(8), 2621–2631. https://doi-org.lili.idm.oclc.org/10.1007/s00394-017-1454-7

Hauser, M. D. (2006). Moral minds: how nature designed our universal sense of right and wrong. Ecco.

Heep, S. (2020). The Long Way of Political Theology to Religious 'Germanism' or How National Socialism Could be Perceived as Fulfillment of Christianity. *Politics, Religion & Ideology*, *21*(3), 311–336. https://doi-org.lili.idm.oclc.org/10.1080/21567689.2020.1786684

History.com Editors. (2023, Nov 16). *Hinduism*. History. Retrieved Feb 6, 2025. https://www.history.com/topics/religion/hinduism

Hoffman, D. (2019). Do we see reality? *New Scientist*, *243*(3241), 34–37. https://doi-org.lili.idm.oclc.org/10.1016/s0262-4079(19)31434-4

Hoffman, D. D. (1998). Visual Intelligence: how we create what we see. W.W. Norton.

Hoffman, D. D. (2019). The case against reality : why evolution hid the truth from our eyes. W W Norton & Co Inc.

Reference List

Hoffman, D. D. (2024). Spacetime Is Doomed: Time Is an Artifact. *Timing & Time Perception*, *12*(2), 189–191. https://doi-org.lili.idm.oclc.org/10.1163/22134468-bja10096

Holaday, L. W., Howell, B., Thompson, K., Cramer, L., & Wang, E. A. (2021). Association of census tract-level incarceration rate and life expectancy in New York State. *Journal of Epidemiology & Community Health*, *75*(10), 1019–1022. https://doi-org.lili.idm.oclc.org/10.1136/jech-2020-216077

Howell, L. D. (2014). Slavery by the Numbers. USA Today Magazine, 142(2826), 49.

Hsu, Andrea. (2022). Thousands of workers were fired over vaccine mandates. For some, the fight goes on. *All Things Considered (NPR)*.

International Business. (2012). New York City's Planned Soda Ban: Paternalism Or Enlightened Leadership? (2012). *International Business Times*.

Isa, A. R., Shuib, R., & Othman, M. S. (1999). The practice of female circumcision among Muslims in Kelantan, Malaysia. Reproductive Health Matters, 7(13), 137–144. https://doi-org.lili.idm.oclc.org/10.1016/S0968-8080(99)90125-8

Jackson, Helen (1994). A Century of Dishonor. United States: Indian Head Books. p. 345.

Reference List

Johnson, K. V.-A., & Steenbergen, L. (2022). Gut Feelings: Vagal Stimulation Reduces Emotional Biases. *Neuroscience, 494*(1), 119–131. https://doi-org.lili.idm.oclc.org/10.1016/j.neuroscience.2022.04.026

Jordan, D. (2003). Wrights and Wrongs? *History Today, 53*(12), 4–5.

Jung, N., Wranke, C., Hamburger, K., & Knauff, M. (2014). How emotions affect logical reasoning: evidence from experiments with mood-manipulated participants, spider phobics, and people with exam anxiety. *Frontiers in psychology, 5*, 570. https://doi.org/10.3389/fpsyg.2014.00570

Kamber, M. (2001). THREE VIEWS OF ONE WAR: talking jihad. *Village Voice, 46*(45), 48.

Kan, E., Knowles, A., Peniche, M., Frick, P. J., Steinberg, L., & Cauffman, E. (2021). Neighborhood Disorder and Risk-Taking Among Justice-Involved Youth—The Mediating Role of Life Expectancy. *Journal of Research on Adolescence (Wiley-Blackwell), 31*(2), 282–298. https://doi-org.lili.idm.oclc.org/10.1111/jora.12596

Karabey, B. (2010). Cry of Nature: A Plant Interfaces with the World Using Music. *Leonardo, 43*(3), 310–311.

Keith, T. (2016, May 23). Evolution Or Expediency? Clinton's Changing Positions Over A Long Career. Morning Edition (NPR).

Reference List

Ken Tanaka, G., Russell, T. A., Bittencourt, J., Marinho, V., Teixeira, S., Hugo Bastos, V., Gongora, M., Ramim, M., Budde, H., Aprigio, D., Fernando Basile, L., Cagy, M., Ribeiro, P., Gupta, D. S., & Velasques, B. (2022). Open monitoring meditation alters the EEG gamma coherence in experts meditators: The expert practice exhibit greater right intra-hemispheric functional coupling. *Consciousness and Cognition*, *102*, 103354. https://doi-org.lili.idm.oclc.org/10.1016/j.concog.2022.103354

Khait, I., Lewin-Epstein, O., Sharon, R., Saban, K., Goldstein, R., Anikster, Y., Zeron, Y., Agassy, C., Nizan, S., Sharabi, G., Perelman, R., Boonman, A., Sade, N., Yovel, Y., & Hadany, L. (2023). Sounds emitted by plants under stress are airborne and informative. *Cell*, *186*(7), 1328.

Klepper, D., Peltz, J. (2014, Jun 6). *Drink up, NYC: Ban on big sodas canned*. Associated Press.

Kocabaş, A., & Çelik, A. (2024). Gazzâlfnin Mûcizeye Yaklaşımının Tehâfütü'l-Felâsife BağlamındaKuantum Kuramı Açısından Değerlendirilmesi. *KADER*, *22*(2), 312–333. https://doi-org.lili.idm.oclc.org/10.18317/kaderdergi.1519364

Korotayev, A. V., Issaev, L. M., & Shishkina, A. R. (2015). Female Labor Force Participation Rate,Islam, and Arab Culture in Cross-Cultural Perspective. Cross-Cultural Research, 49(1), 3–19. https://doi-org.lili.idm.oclc.org/10.1177/1069397114536126

Krakauer, J. (2003). Under the banner of heaven: a story of violent faith. Doubleday.

Reference List

Kuhn, R.L. [Closer to Truth]. (2021, Jul 5). Nima Arkani-Hamed - How Can Space and Time be the Same Thing? [Video] YouTube.
https://www.youtube.com/watch?v=joeDff7EnAU

Lemley, M. A. (2012). The Myth of the Sole Inventor. *Michigan Law Review*, *110*(5), 709–760.

Levin, J. (2020). Hacking the Akashic Records: The Next Domain for Military Intelligence Operations? *World Futures: The Journal of General Evolution*, *76*(2), 102–117. https://doi-org.lili.idm.oclc.org/10.1080/02604027.2019.1703159

Loneman, S. M. R.N. (2024, Mar 13). Blood sugar levels can fluctuate for many reasons. *Mayo Clinic.* https://www.mayoclinic.org/diseases-conditions/diabetes/expert-answers/glucose-levels/faq-20424316#:~:text=My%20blood%20sugar%20level%20is,What%20could%20be%20the%20cause?&text=Many%20factors%20can%20cause%20high,Hormonal%20changes

Maharaj, T., & Winkler, I. T. (2022). "You don't just do it because someone else said so": Menstrual practices and women's agency in the Hindu diaspora of Trinidad. *Culture, Health & Sexuality*, *24*(6), 827–841. https://doi-org.lili.idm.oclc.org/10.1080/13691058.2021.1887938

Machanda, Z. (2018, Jan 3). Chimpanzee Behavior and the Evolution of Human Warfare – Zarin Machanda. *TuftsAlumnni.* [Video].
https://www.youtube.com/watch?v=Q-J0GXYF7HI.

Martinez, M. & Watts, A. (2015, Jun 30). California governor signs vaccine bill that bans personal, religious exemptions. *CNN Health*. June 30, 2015 https://www.cnn.com/2015/06/30/health/california-vaccine-bill/index.html

Matthews, M. (2014, Apr 23). You'll Stop Worrying About Sugar After Reading This Article. *Legion Athletics*. Retrieved Feb 7, 2025. https://legionathletics.com/sugar-facts/?srsltid=AfmBOoqhKfPx6y8ZJUdgdxfA_WnjR7EWIw6tvvDXggzTHjkcliT-BQOh

Merton, R. K. (1961). Singletons and Multiples in Scientific Discovery: A Chapter in the Sociology of Science. *Proceedings of the American Philosophical Society*, *105*(5), 470–486. http://www.jstor.org/stable/985546

Miller, Stephen (May 1968). "The Predatory Behavior of Dileptus Anser". Journal of Protozoology. 15 (2): 313–19.

Mills, D. B., & Canfield, D. E. (2017). A trophic framework for animal origins. *Geobiology*, *15*(2), 197–210.

Mishra, R. C., Ghosh, R., & Bae, H. (2016). Plant acoustics: in the search of a sound mechanism for sound signaling in plants. *Journal of Experimental Botany*, *67*(15), 4483–4494. https://doi-org.lili.idm.oclc.org/10.1093/jxb/erw235.

Mitton, S. (2006). Astroparticle physics and cosmology. *The Lancet*, *367*(9523), 1692–1697.https://doi-org.lili.idm.oclc.org/10.1016/S0140-6736(06)68738-2

Reference List

Mohanta, T. K. (2018). Sound Wave in Plant Growth Regulation: A Review of Potential Biotechnological Applications. *JAPS: Journal of Animal & Plant Sciences*, *28*(2), 1–9.

Montagu, A. (1986). Touching : the human significance of the skin. Perennial Library.

Moran, R. (2022). Wawa may exclude Philly from future expansion due to crime concerns, city councilmember says. In *Philadelphia Inquirer, The (PA)*.

Morrison, R. E., Dunn, J. C., Illera, G., Walsh, P. D., & Bermejo, M. (2020). Western gorilla space use suggests territoriality. Scientific Reports, 10(1), 1–8.

Moscovice, L. R., Hohmann, G., Trumble, B. C., Fruth, B., & Jaeggi, A. V. (2022). Dominance or Tolerance? Causes and consequences of a period of increased intercommunity encounters among bonobos (Pan paniscus) at LuiKotale. *International Journal of Primatology, 43*(3), 434–459. https://doi-org.lili.idm.oclc.org/10.1007/s10764-022-00286-y

Muller, D.A. [Veritasium]. (2020, Jan 29). This equation will change how you see the world (the logistic map). [Video]. YouTube. https://www.youtube.com/watch?v=ovJcsL7vyrk

Muller, D.A. [Veritasium]. (2019, Dec 6). Chaos: The Science of the Butterfly Effect. [Video]. YouTube. https://www.youtube.com/watch?v=fDek6cYijxI

Reference List

Murray, C. H., & Srinivasa-Desikan, B. (2022). The altered state of consciousness induced by Δ9-THC. *Consciousness and Cognition, 102*(1). https://doi-org.lili.idm.oclc.org/10.1016/j.concog.2022.103357

Murray, Peter. [Brusspup]. (2013, June 6). Amazing Resonance Experiment! [Video]. YouTube.https://www.youtube.com/watch?v=wvJAgrUBF4w

Muskopf, Shannan, M.S. *The Lesson of the Kaibab*. Biology Corner. Retrieved Feb 3, 2025. https://www.biologycorner.com/worksheets/kaibab.html#google_vignette

National Archives. (2008). Vietnam War U.S. Military Fatal Casualty Statistics. *Military records.* Retrieved Feb 8, 2025. https://www.archives.gov/research/military/vietnam-war/casualty-statistics

Neitzel, S., Welzer, H. (2012). Soldaten - On Fighting, Killing and Dying: The Secret Second World War Tapes of German POWs. United Kingdom: Simon & Schuster UK.

Nyathi, S., Karpel, H. C., Sainani, K. L., Maldonado, Y., Hotez, P. J., Bendavid, E., & Lo, N. C. (2019). The 2016 California policy to eliminate nonmedical vaccine exemptions and changes in vaccine coverage: An empirical policy analysis. *PLoS Medicine, 16*(12), N.PAG. https://doi-org.lili.idm.oclc.org/10.1371/journal.pmed.1002994

Reference List

O'Brien, G. (2023). Denmark Vesey's Bible: The Thwarted Revolt that Put Slavery and Scriptureon Trial. *Journal of Religious History*, *47*(4), 628–630. https://doi-org.lili.idm.oclc.org/10.1111/1467-9809.12983

Onion, A., Sullivan, M., Mullen, M., & Zapata, C. (2012). Sandy Hook School Shooting. *History.com*. Retrieved Feb 2, 2025. https://www.history.com/this-day-in-history/gunman-kills-students-and-adults-at-newtown-connecticut-elementary-school

Pandit, S., Pradhan, G., Balashov, H., & Schaik, C. (2016). The Conditions Favoring Between-Community Raiding in Chimpanzees, Bonobos, and Human Foragers. Human Nature, 27(2), 141–159. https://doi-org.lili.idm.oclc.org/10.1007/s12110-015-9252-5

Parikh, A., & Miller, C. (2019). Holy Cow! Beef Ban, Political Technologies, and Brahmanical Supremacy in Modi's India. *ACME: An International Journal for Critical Geographies*, *18*(4), 835–874.

Parsons, L. M., Sergent, J., Hodges, D. A., & Fox, P. T. (2005). The brain basis of pianoperformance. *Neuropsychologia*, *43*(2), 199–215. https://doi-org.lili.idm.oclc.org/10.1016/j.neuropsychologia.2004.11.007

Reference List

Perelló, M., Cornejo, M. P., De Francesco, P. N., Fernandez, G., Gautron, L., & Valdivia, L. S. (2022). The controversial role of the vagus nerve in mediating ghrelin's actions: gut feelings and beyond. *IBRO Neuroscience Reports*, *12*(1), 228–239. https://doi-org.lili.idm.oclc.org/10.1016/j.ibneur.2022.03.003

Poundstone, W. (2013). The recursive universe: cosmic complexity and the limits of scientific knowledge. Dover Pubns.

Report on Kaibab deer investigating committee, 1924. (1924). [Place of publication not identified], Oct. 1, 1924.

Rickert-Hall, J. (2020). Waterloo Region became home for slaves who risked everything. *Record, The (Kitchener/Cambridge/Waterloo, ON)*.

Riolo, R. L., Cohen, M. D., & Axelrod, R. (2001). Evolution of cooperation without reciprocity. *Nature*, *414*(6862), 441–443. https://doi-org.lili.idm.oclc.org/10.1038/35106555

Robert, D. (2023). Plant bioacoustics: The sound expression of stress. *Cell*, *186*(7), 1307–1308.

Rodgers, P. (2005). Double-slit effect seen over time too. *New Scientist*, *185*(2489), 14.

Roy, A. (2017). The experiment of Michelson and Morley. *Resonance*, *22*(7), 633–643. https://doi-org.lili.idm.oclc.org/10.1007/s12045-017-0508-8

Ryan, H. (2024). Superfood: Beans. *Life Extension*, *30*(5), 89–90.

Reference List

Saltzman, J. (2017). Desire nothing: Nirvana is nowhere. *International Communication of Chinese Culture, 4*(1), 117–123. https://doi-org.lili.idm.oclc.org/10.1007/s40636-017-0084-3

Sánchez-Cañizares, J. (2019). Classicality First: Why Zurek's Existential Interpretation of Quantum Mechanics Implies Copenhagen. *Foundations of Science, 24*(2), 275–285. https://doi-org.lili.idm.oclc.org/10.1007/s10699-018-9574-y

Saxena, N. (2017). J. M. Coetzee's Aesthetics of Ahimsa: Towards a Gandhian Reading of The Lives of Animals and Disgrace. *English in Africa, 44*(2), 117–141. https://doi-org.lili.idm.oclc.org/10.4314/eia.v44i2.5

Schmidhuber, J. (2009). Ultimate Cognition à la Gödel. *Cognitive Computation, 1*(2), 177–193.

Skoe, E., & Kraus, N. (2012). A little goes a long way: how the adult brain is shaped by musical training in childhood. *The Journal of Neuroscience : The Official Journal of the Society for Neuroscience, 32*(34), 11507–11510. https://doi-org.lili.idm.oclc.org/10.1523/JNEUROSCI.1949-12.2012

Scripp, L., Ulibarri, D., & Flax, R. (2013). Thinking Beyond the Myths and Misconceptions of Talent: Creating Music Education Policy that Advances Music's Essential Contribution to Twenty-First-Century Teaching and Learning. *Arts Education Policy Review, 114*(2), 54–102. https://doi-org.lili.idm.oclc.org/10.1080/10632913.2013.769825

Shaffern, R. W. (1998). Indulgences and saintly devotionalisms in the middle ages. *Catholic Historical Review, 84*(4), 643. https://doi-org.lili.idm.oclc.org/10.1353/cat.1998.0223

Shaheed, F. (1986). The cultural articulation of patriarchy: legal systems, Islam and women. *South Asia Bulletin*, *6*(1), 38–44. https://doi-org.lili.idm.oclc.org/10.1215/07323867-6-1-38

Sheldrake, R., & Smart, P. (2023). Directional Scopaesthesia and Its Implications for Theories ofVision. *Journal of Scientific Exploration*, *37*(3), 312–329. https://doi-org.lili.idm.oclc.org/10.31275/20232897

Sherman, A. (2015, Jun 17). Hillary Clinton's changing position on same-sex marriage. Politifact. Retrieved Feb 9, 2025. https://www.politifact.com/factchecks/2015/jun/17/hillary-clinton/hillary-clinton-change-position-same-sex-marriage/

Shermer, M. (2015). Perception Deception. *Scientific American*, *313*(5), 75. https://doi-org.lili.idm.oclc.org/10.1038/scientificamerican1115-75

Sosin, A., & Neubauer, A. B. (2024). Why we do what we do matters for how we feel: Links among autonomous goal regulation, need fulfillment, and well-being in daily life. *Journal of Personality and Social Psychology*, *127*(5), 1103–1125. https://doi-org.lili.idm.oclc.org/10.1037/pspp0000522

Starkey, N. Ph.D. (2008). Dump your sugar bowl. *Women's Health*, *5*(8), 40. Retrieved Feb 7, 2025. https://research-ebsco-com.lili.idm.oclc.org/c/zgjz6l/viewer/html/kdpotgtjqj?route=details

Stephanopoulos, G. (2024). How the Oj Simpson Trial Changed America. *Good Morning America (ABC)*, 1.

Reference List

Stevenson, I. M.D. (1975). Cases of the reincarnation type. I. Ten cases in India. Charlottesville: University Press of Virginia. Stevenson, I. (1983).

Stevenson, I. M.D. (1993). Birthmarks and Birth Defects Corresponding to Wounds on Deceased Persons. *Journal of Scientific Exploration, Vol. 7, No. 4, pp. 403-410.* https://med.virginia.edu/perceptual-studies/wp-content/uploads/sites/360/2016/12/STE39stevenson-1.pdf

Stevenson, I. M.D. (1983). American Children Who Claim to remember Past Lives. *The Journal of Nervous and mental Disease, Vol. 171, No.12.* Williams & Wilkens Co. https://med.virginia.edu/perceptual-studies/wp-content/uploads/sites/360/2016/12/STE17.pdf

Strickland, S. J., & Burriss, K. G. (2001). Music and the Brain in Childhood Development. *Childhood Education,* 78(2), 100.

Strickland, S. J. (2002). Music and the Brain in Childhood Development. Review of Research. In *Childhood Education* (Vol. 78, Issue 2, pp. 100–103).

Suomi, S. J., Harlow, H. F., & Kimball, S. D. (1971). Behavioral Effects of Prolonged Partial Social Isolation in the Rhesus Monkey. *Psychological Reports*, *29*(Supplement 3), 1171–1177. https://doi-org.lili.idm.oclc.org/10.2466/pr0.1971.29.3f.1171

Surles, Ashlea. (2016). Mountain Charity Inspired by Letters from Mother Teresa. *ABC13News.* https://wlos.com/news/local/mountain-charity-inspired-by-letters-from-mother-teresa

Tamis-LeMonda, C. S., & Masek, L. R. (2023). Embodied and Embedded Learning: Child, Caregiver, and Context. *Current Directions in Psychological Science*, *32*(5), 369–378. https://doi-org.lili.idm.oclc.org/10.1177/09637214231178731

Tanaphong Uthayaratana, Nattasuda Taephant, & Kullaya Pisitsungkagarn. (2019). Four noble truths based problem solving: a therapeutic view. *Mental Health, Religion & Culture*, *22*(2), 119–129. https://doi-org.lili.idm.oclc.org/10.1080/13674676.2018.1512565

Tomasello, M., & Vaish, A. (2013). Origins of human cooperation and morality. *Annual Review of Psychology*, *64*, 231–255. https://doi-org.lili.idm.oclc.org/10.1146/annurev-psych-113011-143812

Tomkins, A., Duff, J., Fitzgibbon, A., Karam, A., Mills, E. J., Munnings, K., Smith, S., Seshadri, S. R.,Steinberg, A., Vitillo, R., & Yugi, P. (2015). Controversies in faith and health care. *The Lancet*, *386*(10005), 1776–1785. https://doi-org.lili.idm.oclc.org/10.1016/S0140-6736(15)60252-5

Travis, F. (2014). Transcendental experiences during meditation practice. *Annals of the New York Academy of Sciences*, *1307*, 1–8. https://doi-org.lili.idm.oclc.org/10.1111/nyas.12316

Tucker, J. B. (2005). Life before life : a scientific investigation of children's memories of previous lives. St. Martin's Press.

Tucker, J. B. (2008). Life before life : children's memories of previous lives. St. Martin's Griffin.

Reference List

Tufts Health Letter. Pick Strawberries to Benefit Your Heart and Brain. (2013). *Tufts University Health & Nutrition Letter, 31*(4), 6.

Tucker, J.B., Greyson, B., Kelly, E.F., Penberthy J.K. (2017, Mar 7). Is There Life after Death? Fifty Years of Research at UVA. *UVA Medical Center Hour.* [Video]. YouTube. https://www.youtube.com/watch?v=0AtTM9hgCDw

Venn, J. (1962). *The logic of chance*. Chelsea Publications.

Victor, B. (2003). Army of Roses: inside the world of Palestinian women suicide bombers. Rodale.

Vinx, L. (2021). Carl Schmitt and the authoritarian subversion of democracy. *Philosophy & Social Criticism, 47*(2), 173–177. https://doi-org.lili.idm.oclc.org/10.1177/0191453720974724

Wang, H., Li, T., Bezerianos, A., Huang, H., He, Y., & Chen, P. (2019). The control of a virtual automatic car based on multiple patterns of motor imagery BCI. *Medical & Biological Engineering & Computing, 57*(1), 299–309. https://doi-org.lili.idm.oclc.org/10.1007/s11517-018-1883-3

Weizmann Institute Of Science. (1998, February 27). Quantum Theory Demonstrated: Observation Affects Reality. *ScienceDaily*. Retrieved February 24, 2025 from www.sciencedaily.com/releases/1998/02/980227055013.htm.

Reference List

Westcott, T. (2016). Torture, starvation, deprivation: life inside IS prisons in Libya. *The New Humanitarian.* Retrieved Feb 5, 2025. https://www.thenewhumanitarian.org/feature/2016/10/24/torture-starvation-deprivation-life-inside-prisons-libya

Williams, Matt. (2017, July 3). Universe Today Space and Astronomy News. Retrieved Feb 1, 2025. https://www.universetoday.com/45047/how-far-does-light-travel-in-a-year-1/#:~:text=But%20how%20far%20does%20light,5%2C878.5%20billion%20miles)%20per%20year.

Wintrobe, R. (2001). How to understand, and deal with dictatorship: an economist's view. *Economics of Governance, 2*(1), 35–58. https://doi-org.lili.idm.oclc.org/10.1007/s10101-001-8001-x

Yu, Y., Zhou, Z., Yin, E., Jiang, J., Tang, J., Liu, Y., & Hu, D. (2016). Toward brain-actuated car applications: Self-paced control with a motor imagery-based brain-computer interface. Computers in Biology and Medicine, 77(1), 148–155.

Yun, G., Jiali, X., Yun, G., & Jiali, X. (2024). Research and exploration on cymatics in sound visualization. *E3S Web of Conferences, 486*(1). https://doi-org.lili.idm.oclc.org/10.1051/e3sconf/202448603002

Zatorre, R., & McGill, J. (2005). Music, the food of neuroscience? *Nature, 434*(7031), 312–315. https://doi-org.lili.idm.oclc.org/10.1038/434312a

Reference List

Zeller, B. (2023, Aug 23). How Many Died in the American Civil War? *History*. Retrieved Feb 7, 2025. *https://www.history.com/news/american-civil-war-deaths*

Zelman, K.M. (2024, Feb 12). What's the difference Between Sucrose and Fructose? *WebMD*. Retrieved Mar 17, 2025. https://www.webmd.com/diet/whats-the-difference-between-sucrose-and-fructose

Zhubi-Bakija, F., Bajraktari, G., Bytyçi, I., Mikhailidis, D. P., Henein, M. Y., Latkovskis, G., Rexhaj, Z., Zhubi, E., Banach, M., Alnouri, F., Amar, F., Atanasov, A. G., Bartlomiejczyk, M. A., Bjelakovic, B., Bruckert, E., Cafferata, A., Ceska, R., Cicero, A. F. G., Collet, X., ... Zirlik, A. (2021). The impact of type of dietary protein, animal versus vegetable, in modifying cardiometabolic risk factors: A position paper from the International Lipid Expert Panel (ILEP). *Clinical Nutrition, 40(1)*, 255–276.

Zielinski, B. (2022). Théories socio-économiques et critiques de la démocratie parlementaire: Carl Schmitt et son influence. *Revue d'Allemagne et Des Pays de Langue Allemande, 54(2)*, 413–427. https://doi-org.lili.idm.oclc.org/10.4000/allemagne.3257

Author Biography

I was raised just outside of Philadelphia, Pennsylvania then attended music college in Boston, Massachusetts, at Berklee College of Music for two years. After working construction for a few months, I went back to college at Pennsylvania State University and graduated in 1990. I met my wife Carolyn at Penn State, and we moved out to Idaho after graduation. I worked as a surveyor and fire fighter (seasonally) in the USDA Forest Service from 1990-2008. In 1996, I earned my teaching credential from Lewis & Clark State University. I was a regular classroom teacher in public school from 1997-2008, when I earned my master's degree in Education Administration. I was a principal at all levels (elementary, middle, junior high, and high school) in public school from 2008-2024. I dealt with everyday troubles, sometimes horrendous situations, and also incredible triumphs, with thousands of young people for nearly thirty years. Counseling, helping, making decisions with young people who were grappling with finding meaning, is one of the reasons I felt capable to write this book. My own family consisting of my wife and five children, were always questioning big ideas such as truth, and meaning. Getting older, dealing with the death of close loved ones, sent me on a research quest, to base conclusions on scientific study and logical reason. This book is a hodge podge of ideas written down (about fifteen years in the making) as I encountered challenging situations and ideas. I am a musician, a cook, I travel incessantly (love immersion in other cultures) and I stay fit, even into my older years. I previously had an article published through the Idaho Association of School Administrators which centered on the importance of teaching music.

www.ingramcontent.com/pod-product-compliance
Lightning Source LLC
Chambersburg PA
CBHW070849050426
42453CB00012B/2101

CAREER LESSONS